Essential Concepts of Oceanography

Essential Concepts of Oceanography

Edited by **Theodore Roa**

New York

Published by Callisto Reference,
106 Park Avenue, Suite 200,
New York, NY 10016, USA
www.callistoreference.com

Essential Concepts of Oceanography
Edited by Theodore Roa

International Standard Book Number: 978-1-63239-316-6 (Hardback)

Contents

Preface   VII

Chapter 1   **Stable Isotope Methods for the Study of the Nitrogen Cycle**   1
Evgenia Ryabenko

Chapter 2   **A Statistical Approach for Wave-Height Forecast Based on Spatiotemporal Variation of Surface Wind**   41
Tsukasa Hokimoto

Chapter 3   **Challenges and New Advances in Ocean Color Remote Sensing of Coastal Waters**   60
Hubert Loisel, Vincent Vantrepotte, Cédric Jamet and Dinh Ngoc Dat

Chapter 4   **Near Surface Turbulence and Gas Exchange Across the Air-Sea Interface**   98
Qian Liao and Binbin Wang

Chapter 5   **Novel Tools for the Evaluation of the Health Status of Coral Reefs Ecosystems and for the Prediction of Their Biodiversity in the Face of Climatic Changes**   125
Stéphane La Barre

Permissions

List of Contributors

Preface

This book aims to highlight the current researches and provides a platform to further the scope of innovations in this area. This book is a product of the combined efforts of many researchers and scientists, after going through thorough studies and analysis from different parts of the world. The objective of this book is to provide the readers with the latest information of the field.

The essential concepts of oceanography are descriptively discussed in this advanced book. Oceanography is a quintessential interdisciplinary science and because of its unique setting inside a fluid environment, it makes connections highly effective. These ocean connections have been well elucidated in this book and it reflects a quite explicit multi-environmental and multi-disciplinary character. This book comprises of information regarding distinct topics under extremely distinct settings, with both focused as well as broad-view approaches. It lays stress on the idea that there is a need to comprehend the various minor perspectives in order to understand the bigger picture of the overall mechanism of the ocean.

I would like to express my sincere thanks to the authors for their dedicated efforts in the completion of this book. I acknowledge the efforts of the publisher for providing constant support. Lastly, I would like to thank my family for their support in all academic endeavors.

Editor

Stable Isotope Methods for the Study of the Nitrogen Cycle

Evgenia Ryabenko

Additional information is available at the end of the chapter

1. Introduction

Nitrogen, a limiting element for biological productivity, plays a key role in regulating the biogeochemical processes in the ocean. In today's ocean, all of the major reactions in the N cycle are mediated by assimilatory or dissimilatory functions of marine organisms. Because marine organisms preferentially incorporate lighter stable isotope of nitrogen ^{14}N instead of ^{15}N, each major metabolic reaction in the N cycle involves irreversible kinetic fractionation of nitrogen.

The isotopic composition of a pool of nitrogen can be used to identify the relative importance of sources that are isotopically distinct, or processes that add or remove nitrogen with a characteristic pattern of isotopic discrimination. The strongest isotopic fractionations are associated with dissimilatory processes that mediate the transfer of nitrogen from one inorganic pool to another [1]. In contrast, processes such as primary production, which move nitrogen directly into and through the food web are associated with comparatively weak isotopic fractionations [1]. The extent of nitrogen isotope fractionation also depends upon the kinetics of individual metabolic reactions, concentration of products and reactants, environmental conditions (e.g., oxygen concentrations) and the microbial species involved.

The crucial challenge in using nitrogen isotope methods is the complexity of the marine nitrogen cycle and the potential influence of multiple processes on the isotopic Composition of several biologically active pools of nitrogen. Successful use of nitrogen isotopes in resolving N cycle fluxes and processes requires an understanding of the general distribution of nitrogen isotopes in marine systems, the nature of isotopic fractionation, and a careful consideration of the processes at work in the system of interest. In this chapter I will introduce the reader to the two different approaches of nitrogen isotope analysis:

natural abundance or tracer methods (^{15}N-labeling). The natural abundance of the stable isotope of nitrogen can provide critical insights into processes acting on a variety of scales with minimal alteration and manipulation of the system being studied. The isotopic signal collected in this type of approach is an integrated value that incorporates large spatial and temporal variation of the relevant processes. Conversely, isotope tracer methods involve short-term incubations of small, isolated samples of water under in situ or simulated in situ conditions. Tracer methods are primarily used for rate experiments that complement larger-scale natural abundance studies.

I begin the chapter with basic definitions of isotopic fractionation and analytical considerations of the sample measurements. I follow with an overview of the processes of nitrogen sources and sinks. I conclude with a discussion of how isotopic data can contribute to the current debate on the balance of N-inputs and losses in the nitrogen cycle.

2. Fundamentals

2.1. Isotopes and calculation of their ratio

Isotopes are atoms of an element that share the same number of protons but a different number of neutrons. In the scientific nomenclature, isotopes are specified in the form $_{n}^{m}E$, where "m" indicates the mass number (the sum of protons and neutrons in the nucleus) and "n" refers to the atomic number of an element "E". There are more than 10 nitrogen isotopes known. Most of these isotopes are radioactive and highly unstable with longest half-life time for $_{7}^{13}N$ of 10 minutes. The only two stable nitrogen isotopes are $_{7}^{14}N$ and $_{7}^{15}N$, which have seven protons each and seven or eight neutrons in their nucleus, respectively. ^{15}N is the less frequent stable isotope, constituting of 0.365% of the global nitrogen pool [2]. Consequently, it is more practical to measure the difference or ratio of two isotopes instead of the absolute quantity of each.

Isotopic compositions are expressed in terms of "delta" (δ) values which are given in parts per thousand or per mil (‰). Nitrogen isotope ratios, for example, are expressed as the ‰-difference to atmospheric N_2, which has a constant $^{14}N/^{15}N$ of 272 ± 0.3 [3]. The $\delta^{15}N$-value in the sample is then calculated by the following equation (1).

$$\delta^{15}N(vs.air) = \left(\frac{(^{15}N/^{14}N)_{sample}}{(^{15}N/^{14}N)_{air}} - 1 \right) \times 1000 \qquad (1)$$

The δ-values do not represent absolute isotope abundances but rather the ‰-difference to a widely used reference standard, such as VSMOW (Vienna Standard Mean Ocean Water). The δ-value is then calculated from equation (2), by measuring the isotope ratios (R) for the sample and the reference standard:

$$\delta^{15}N = \left(\frac{R_{sample}}{R_{standard}} - 1 \right) \times 1000 \tag{2}$$

where R_{sample} and $R_{standard}$ represent the isotope ratio $\delta^{15}N_{(vs.air)}$ in the sample and in the standard respectively, calculated using equation (1). By convention, R is the ratio of the less abundant isotope over the most abundant isotope (i.e. $^{15}N/^{14}N$ for nitrogen).

As previously discussed, the $\delta^{15}N$ value changes under the influence of chemical and physical processes. If the process is complete the resulting $\delta^{15}N$ value in the product is equal to the value of the reagent. Only if the reaction is incomplete the fractionation of isotopes happens, which means that the $\delta^{15}N$ values in the product and substrate differ.

2.2. Isotope fractionation effect

There are two different fractionation processes, both of which will be discussed here: equilibrium and kinetic fractionation processes.

Equilibrium fractionation processes are reversible processes. They are mainly driven by changes in the internal energy of a molecule, i.e. vibrations of the atoms within a molecule with respect to each other and rotations around the molecular axes. The equilibrium fractionation factor $\alpha_{eq.}$ is related to the equilibrium constant K as shown in equation (3), where n is the number of exchanged atoms.

$$\alpha_{eq.} = k^{1/n} \tag{3}$$

During equilibrium reactions, the heavier isotope preferentially accumulates in the compounds with a higher number of bonds. During phase changes, the ratio of heavy to light isotopes in the molecules also changes. For example, as water vapor condenses in rain clouds (a process typically viewed as an equilibrium process), the heavier water isotopes (^{18}O and ^2H) become enriched in the liquid phase while the lighter isotopes (^{16}O and ^1H) remain in the vapor phase. In addition, the equilibrium isotopic effect decreases as the system temperature increases.

Kinetic fractionation processes are also associated with incomplete processes like evaporation, dissociation reactions, biologically mediated reactions and diffusion [4]. Kinetic isotope fractionation reflects the difference in the bond strengths or motilities of the isotopic species. The degree of isotopic fractionation associated with a reaction is commonly expressed with α, which is the ratio of rate constants for molecules containing the different isotopes:

$$\alpha = {}^{14}k / {}^{15}k \tag{4}$$

where ^{14}k and ^{15}k are the rate constants for molecules containing the light and heavy isotopes, respectively. Most biological reactions discriminate against the heavier isotope, yielding

1.00$<\alpha<$1.03. This makes it convenient to define an isotopic enrichment factor (ε) that highlights more clearly the range of variation:

$$\varepsilon = (\alpha - 1) \times 1000 \qquad (5)$$

An alternative method for describing isotopic enrichment is the Rayleigh relationship. This is an exponential relationship that describes the partitioning of isotopes between two reservoirs as one reservoir decreases in size. The equation can be used if the following conditions are met: 1) material is continuously removed from a mixed system; 2) the fractionation accompanying the removal process at any instance is described by the fractionation factor α, and 3) α does not change during the process. Under these conditions, the evolution of the isotopic composition in the residual (reactant) material is described by:

$$R = R_0 f^{(\alpha-1)}, \qquad (6)$$

where R is the isotopic ratio of the product, R_0 is the initial ratio of the reactant, f is the fraction of the substrate pool remaining and α is the kinetic fractionation factor.

2.3. Measurements of δ^{15}N

Stable isotope measurements are becoming a routine tool in studies of marine ecosystems due to the increasing availability of mass spectrometers. Biological oceanographers usually using continuous flow systems for ^{15}N measurements that integrate a preparatory system (e.g., an elemental analyzer) with a mass spectrometer [5]. The mass spectrometer typically measures the isotopic composition of N_2 generated by combustion and carried in a stream of He gas to an open split interface that introduces a small fraction of the gas stream into the ion source for measurement. This analytical approach requires little or no manipulation of organic samples, though care must be taken to avoid contamination with exogenous nitrogen, particularly when tracer-level ^{15}N experiments are being carried out in the vicinity. Since N_2 is typically the analyte, minimizing atmospheric contamination in the preparatory system is also quite important. Alternatively, N_2O gas may be used for isotope measurements, which practically eliminates the potential contamination from the atmosphere and gives additional information of oxygen isotope composition of nitrogen compound (e.g nitrate or nitrite). In this case, nitrogen compounds must be chemically converted into N_2O prior the isotope analysis [6-11]. Special care has to be taken due to the potential exchange of oxygen isotopes between nitrate and water leading to deviation of initial value (e.g. [10]).

Tracer-level ^{15}N experiments used for rate measurements typically add about 10% of the ambient source concentration of ^{15}N-compound. The initial planning of tracer studies requires consideration of several practical issues. First of all, one has to determine the length of incubation, which is connected to turnover rate of the relevant pools and the detection limits for the analytical methods. The turnover rate (pool size$^{-1}\times$uptake rate) has to be estimated for both source and product pools to ensure that tracer can be detected in the product pool, while dilution of the source pool is either insignificant or small [12, 13]. Increasing incubation time will increase the amount of tracer in the product but it will reduce the potential of isotope dilution affecting the source pool. Given a best-guess turnover rate for the product pool, one can estimate the relative trade-offs between incubation time and changes in isotopic labeling of the source and product pools at a given enrichment level (7).

Turnover rate × incubation time × source enrichment = change in enrichment, e.g., $0.1 \ d^{-1} \times 0.2 \ d^{-1} \times 10\% = 0.2\%$.   (7)

Generally the shortest incubation possible would resolve this dichotomy for biogeochemical processes such as nitrification or denitrification. However, complications may arise for uptake studies since short incubations primarily represent the transport (movement across the cell membrane) of N-compounds, whereas longer incubations primarily represent assimilation into organic compounds such as amino acids or proteins.

Finally, one has to consider the degree of replication and treatment of the initial, or time zero, concentrations and enrichment of both the product and source pools. To account for small, natural (%) level changes in these pools, studies either assume that particles have constant natural abundance values, or they assess the natural abundance periodically during the study. Alternatively, tracer can be added to the incubation bottle, which is then deployed and then immediately filtered at an initial time point for comparison with the final values. One then needs to decide how to interpret the initial time point, as there is often a substantial time between tracer addition and filtration (often 30 min). During this interval, the population may experience abnormal conditions (e.g., high light exposure or rush uptake due to nutrient perturbation) prior to filtration. It was also shown recently by Mohr *et al.* that measurements of N_2-fixation rates by introducing a tracer in bubble form can lead to significant underestimation of the process [14]. The equilibration, i.e. the isotopic exchange between the $^{15}N_2$ gas bubble and the surrounding water is controlled primarily by diffusive processes. The major variables that influence the rate of isotopic exchange include the surface area to volume ratio of the bubble, the characteristics of the organic coating on the bubble surface [15], and temperature and the rate of renewal of the water-bubble interface [16]. To avoid this problem, the authors recommend adding an aliquot of $^{15}N_2$-enriched water to incubators used in rate measurements.

Samples of oceanographic interest often pose additional analytical challenges simply because they are difficult to obtain in large quantities and most mass spectrometry systems require fine-tuning and careful attention to leaks in order to process samples containing less than a few micromoles of nitrogen. In practice, open ocean samples including suspended particles, small zooplankton, and sinking organic matter are often available only in μmole or sub-μmole quantities. These low concentrations create a strong incentive to minimize the mass requirements of the analytical systems used for marine samples, and to develop methods for correcting for the influence of any analytical blank, which will disproportionately affect small samples. Even with substantial care to reduce this source of contamination, a nitrogen blank on the order of 0.05–0.15 μmol N is typical for systems in current use [17]. For analysis of the smallest samples, the influence of this blank can be reduced provided that an appropriate range of mass and isotopic standards is run, allowing estimation of the size and $\delta^{15}N$ of the blank.

Operationally, isotope ratio mass spectrometers measure the difference in ^{15}N abundance between a sample and a reference gas calibrated to atmospheric N_2. This reference gas calibration may be carried out directly by comparison to atmospheric values [18], or indirectly using any of a number of organic and inorganic secondary standards (e.g., nitrate and ammonium salts, acetanilide, glutamic acid) available from NIST or IAEA. For precise system calibration and correction of blank effects, it is also important to analyze a size series of a working standard that is chemically similar to the samples of interest.

Correcting for an analytical blank is a multi-step process involving separate regression analyses to evaluate the size of the blank and the dependence of $\delta^{15}N$ values on size. This information can then be used to remove the isotopic influence of the blank. Any real sample can be treated as a mixture of sample material and contaminants of various origins. If the blank contribution is constant in composition and magnitude across analyses, then a simple mass balance can be constructed for the mixture analyzed:

$$M_{mix}\delta^{15}N_{mix} = M_{sample}\delta^{15}N_{sample} + M_{blank}\delta^{15}N_{blank'} \tag{8}$$

where M is the size of an individual pool (sample, blank, or the mixture). In general, the nitrogen analytical blank is small enough that blank corrections are minor for samples containing more than 1.5–2 µmol of nitrogen.

3. Sampling for $\delta^{15}N$ measurements

Modern water column nitrogen isotope analysis is used on broad variety of samples as particles and dissolved nitrogen species, e.g. nitrate, ammonia or N_2 and N_2O. For each of these methods, special care must be taken during sample collection and preparation.

3.1. Particles

The simplest and most common method to remove particles from incubation water is by filtration. During the filtration special care should be taken to minimize the pressure difference to avoid strong cell rupture [18]. Several studies showed that Teflon and aluminium oxide filters capture up to 60% higher amount of particulate organic nitrogen (PON) than the GF/F filters [19, 20]. GF/F filters have been also found to retain only ~50% of bacteria [21]. In addition to incomplete retention of particles, glass-fiber filters adsorb dissolved organics on the high surface area of the filter [22]. For ^{15}N studies, although mass retention of DON is not a major issue, retention of the highly labeled incubation water on the filter has the potential to affect the isotopic ratio of the retained particulates so that wetting the filter prior to filtration and rinsing afterwards with filtered seawater is a common practice. Filtered seawater is used to prevent osmotic shock of the living particles. Regardless of the type of filter, the particle load needs to be preserved and converted into the gaseous form prior to mass spectroscopic analysis. $\delta^{15}N$ of particles can be analysed directly via combustion (conversion into N_2) or via persulfate oxidation (conversion into NO_3^-) [23].

3.2. Nitrate and nitrite

The most abundant dissolved nitrogen specie in the ocean is nitrate (NO_3^-) followed by nitrite (NO_2^-). In order to perform the isotopic analysis both have to be converted into forms amenable to enter the mass spectrometer. The methods differ by level of reduction, with conversion from NO_3^- to NO_2^-, NO, N_2 or complete reduction to ammonium.

Although NO_2^- concentrations are usually very small compared to NO_3^-, there are cases when both should be analyzed without an interference of $\delta^{15}N$ signals of one another. To do this one has to separate NO_3^- and NO_2^- in the sample by excluding one of them from the sample. Olson [24] used sulfamic acid, while Yakushiji and Kanda [25] took advantage of the reaction of sulfanilic acid with NO_2^-. Rather than forming the final azo dye, the diazonium ion can be destroyed by heating which effectively destroys the initial NO_2^- present in the sample. Two new techniques remove nitrite by reaction with sodium azide in acetic acid buffer to produce N_2O [7] or via reaction with ascorbic acid to produce NO which is then removed bubbling [11].

Nitrate conversion to ammonia commonly uses DeVarda's alloy approach [26] combined with moderate heating. However, this method can add a significant blank that varies with the batch, and combustion of the alloy failed to reduce the blank while also reducing the reduction efficiency [27]. Tanaka and Saino [28] replaced Devarda's alloy with an aluminum reagent that reduced the blank and permitted analysis of nitrate at low concentrations.

Biological reduction of NO_3^- to NO_2^- or gaseous forms has been used by several investigators. Risgaard-Petersen et al. [29] and [30] used cultures of denitrifying bacteria to respire NO_3^- and NO_2^- to N_2 which was then stored for analysis by mass spectrometry. Sigman et al [8] changed this method by using mutant strain of a denitrifying bacteria that only produced N_2O to avoid the problem with contamination of N_2 from the air. The chemical alternative using sodium azide buffer to produce N_2O from NO_2^- was developed by McIlvin and Altabet [7] for natural abundance measurements in seawater samples. It was further improved for samples with low salinities [6] and can be used to analyze nitrate as well as nitrite in the samples.

3.3. Ammonium

Ammonium is not necessarily dominant by mass, but it is often predominant with respect to flux. NH_4^+ is produced primarily by the breakdown and remineralization of organic forms of nitrogen and consumed by autotrophic assimilation and re-incorporation into organic molecules, chemosynthetic nitrification to NO_2^-, and anammox oxidation by NO_2^- to N_2. For isotopic analysis ammonia has to be extracted quantitatively from the solution and converted to suitable gaseous species (e.g. N_2).

Initial approaches to isolating ammonium for isotopic analysis by steam distillation were derived from the soil literature. In all methods involving NH_4^+, great care is required to minimize ambient contamination (e.g., cleaning compounds such as Windex), from sample/ reaction containers and from reagents.

Distillation of NH_4^+ under alkaline conditions with subsequent conversion to dinitrogen gas has been used by several groups but is time consuming and has problems with cross-contamination and fractionation [31-34]. In the diffusion method, ammonia passes through a membrane into an acidic medium and is trapped as an NH_4^+ salt [34-37]. This approach, however, is also time-intensive and not very reliable at low concentrations [34]. For freshwater samples Hg precipitation and cation exchange are reported to show very good results for isotope analysis [32, 38]. An organic reaction with NH_3 producing indophenol permits organic or solid-phase extraction works well for ^{15}N-labeled samples [32] but cannot be used for samples with $[NH_4^+] < 5\mu M$ [38]. Zhang and Altabet [39] developed a robust method for sea water samples at low concentrations (0.5 μM or lower), which involves oxidation of ammonia into NO_2^- and subsequent reduction to N_2O.

3.4. N_2O and N_2

In addition to the previously discussed methods where production of N_2 or N_2O from other forms of nitrogen is an analytical step, there are related methods for analysis of the gases where the gas itself is the form of interest. N_2O can be formed during nitrification and denitrification processes, while N_2 is the end product of denitrification and anammox. Isotope paring technique (IPT) for analysis of N_2 requires analysis of all three isotopic variants of N_2 ($^{28}N_2$, $^{29}N_2$, $^{30}N_2$) [40]. Either a quadrupole or magnetic sector mass spectrometer can be used for the isotope pairing analysis, but the quadrupole can also simultaneously determine the N_2/Ar ratio. Extraction of N_2 from water samples, generally stored in a gas-tight container sealed with a septum (e.g., Exetainer), involves partitioning of the gas into a suitable headspace such as helium. The gas is allowed to equilibrate and the headspace is sampled for introduction of the gas phase into the mass spectrometer [41].

Rather than extraction into a headspace, several investigators have employed membrane inlet mass spectrometers (MIMS) to directly sample N_2 for denitrification and nitrogen fixation studies [30, 42]. MIMS can also be used directly to measure denitrification by monitoring changes in the N_2/Ar ratio over time since the Ar composition reflects changes in temperature and other physical factors while change in the N_2 concentration reflects the balance between nitrogen fixation and denitrification [43]. An advantage of combining MIMS with the isotopes is the ability to separate the two contributions to the net flux. MIMS also obviates the need for the generally slow, separate extraction step. The membrane inlet configuration varies but always features a gas-permeable membrane that separates the mass spectrometer from the solution [44, 45]. A probe inlet [44]permits analysis of depth zonation of rate processes in sediments but has only been employed for N_2/Ar measurements.

Analysis of $^{15}N_2O$ during nitrification and denitrification rate studies has become increasingly popular. Punshon and Moore [46] further developed the method to extract $^{15}N_2O$ produced from either $^{15}NH_4^+$ or $^{15}NO_3^-$ by natural samples. Using purge-trap gas chromatography with a quadrupole mass spectrometer, they were able to determine production rates for N_2O from nitrification. The origin of N_2O (denitrification or nitrification) can be determined using isotopic signature of oxygen isotope. $\delta^{18}O$ isotope signature in N_2O molecule is a product of mixing from dissolved O_2 and H_2O and any fractionation process associated with its production [47]. For instance, production of N_2O via hydroxylamine oxidation (nitrification) will only include oxygen atoms from dissolved O_2, while via denitrification it will have $\delta^{18}O$ isotope of NO_2^- or NO_3^-, originating from both dissolved O_2 and H_2O [47, 48].

The site preference isotopic signature (isotopomers) is also used to identify pathways of N_2O production. In the molecule of N_2O the ^{15}N can be at a central (α) or outer (β) position. The intermolecular distribution of ^{15}N is expressed as site preference (SP) (eq.9)[49].

$$SP = \delta^{15}N^{\alpha} - \delta^{15}N^{\beta} \tag{9}$$

In contrast to $\delta^{18}O$ signature, SP is not dependent on the substrate's isotopic composition but reflects the microbial production mechanism [48]. For example, production of N_2O via hydroxylamine oxidation results in an SP of 33‰, while via denitrification the SP is 0‰ [50]. Good summery on isotopomers for soil studies is presented in the book of Environmental Isotope Geochemistry [51] showing the variable range of SP values for nitrification. In general, isotopomers are valuable tool for distinguishing between nitrification and denitrification processes of N_2O production.

4. Main processes of nitrogen cycle and their influence on isotope balance

Most nitrogen in marine environments is present in five forms: N_2, a quite stable molecule that requires specialized enzymatic systems to break and use it; NO_3^-, the most oxidized form of nitrogen and the dominant form of N within oxic environments; NH_4^+, the most reduced natural form of N and the dominant biologically available form found in anoxic environments; particulate nitrogen, predominant within sediments and primarily in the form of organic N, and dissolved organic N (DON), which is a complex mixture of compounds with a wide range of compositions [52-55]. Nitrate, nitrite, ammonium, and organic nitrogen are typically grouped together as "fixed N" in discussions about the availability of nitrogen. Although each form has a different level of reactivity, a complex web of reactions links these different compounds. The most important processes discussed here are: N_2 fixation, nitrification, assimilation and N-loss (conversion of fixed N to N_2). As it was mentioned already earlier, all biologically mediated processes lead to nitrogen isotope fractionation. For example, denitrification converts NO_3^- to N_2, which has the net effect of depleting the ocean's supply of combined nitrogen and doing so with strong discrimination against ^{15}N. This results in an increase in the $\delta^{15}N$ of the major pool of oceanic combined nitrogen (NO_3^-) concomitant with a reduction in its overall size. In contrast, N_2-fixation adds combined nitrogen with a low $\delta^{15}N$ to the ocean, counteracting both the mass and isotopic effects of denitrification (figure 1). In the subsequent parts of this chapter I will discuss all these processes in detail and their influence on the nitrogen isotope budget in the ocean.

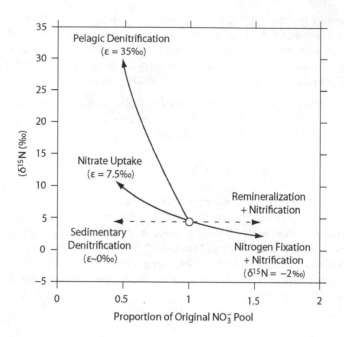

Figure 1. The impact of different processes on the $\delta^{15}N$ of oceanic NO_3^- [56]. Axes show deviation of $\delta^{15}N$ signal from oceanic average and loss/input of nitrogen due to different processes.

4.1. Nitrogen fixation in the ocean

Microbial N_2 fixation in the ocean inputs "new" N into the surface waters by reducing N_2 gas into two molecules of ammonia. The reaction is logically consumes energy and is catalyzed by nitrogenase enzyme complex, which is extremely oxygen sensitive (req. 10).

$$N_2 + 8\,H^+ + 16\,ATP + 8\,e^- \rightarrow 2\,NH_3 + H_2 + 16\,ADP + 16\,P_i \qquad (10)$$

The nitrogenase proteins are highly similar among diazotrophs, and the well-conserved *nifH* gene is commonly used for phylogenetic and ecological studies [57]. In the open ocean N_2 fixation research has focused most intensively on the cyanobacterium, *Trichodesmium* [58, 59], which often occur as aggregates (often referred to as colonies) visible to the naked eye ('sea sawdust'). It can also occur, however, as individual filaments. In early reports N_2 fixation was observed in the pelagic zone and associated with the cyanobacterial epiphytes (*Dichothrix fucicola*) of the brown macroalga, *Sargassum* [60] and cyanobacterial endosymbionts of certain oceanic diatoms [61, 62]. *Trichodesmium* has a unique physiology that employs spatial and temporal segregation and increased oxygen consumption to allow it to simultaneously fix N_2 and CO_2 (and thus evolve O_2)[63].

Evidence of N_2 fixation began to first accumulate in the late 1970s with several pioneering N isotope studies examining the natural abundance of the ratio of $\delta^{15}N$ in the western Pacific Ocean [59, 64]. The average isotopic signature of nitrate in the deep ocean lay about 5‰ [27, 65, 66]. Nitrogen fixation is the process by which N_2 is incorporated from the air into the cell. As the $\delta^{15}N_{air}$ is set to 0‰, the kinetic isotope effect of the process is small (-2‰ to +2‰). Surface particulate and dissolved inorganic N pools in the Kuroshio often had a ^{15}N-depleted isotopic signature, indicative of a N_2 fixation source [67]. Using *nifH* gene studies, N_2-fixing was found also in the picoplankton, as well as in heterotrophic bacteria from the guts of copepods [68]. Heterotrophic N_2 fixation does not consume CO_2 during the process as diazotrophs do but instead produces it. However, the mechanism of heterotrophic N_2 fixation is not yet understood and it is still unclear and the potential contribution of these organisms to marine N_2 fixation is uncertain.

4.1.1. Controlling factors for N_2 fixation and its distribution

Numerous factors, physical, chemical and biotic, can affect the extent of N_2 fixation in an ecosystem [69-72]. Many factors that affect nitrogenase activity co-vary, such as light, temperature, oxygen and turbulence, making it difficult to determine the relative impact of each. Species-level responses to these variables also differ. The heterocystous diazotrophs, for instance, are typically restricted to special marine environments. Metabolically active populations of *Trichodesmium* have been observed at 18.3 °C [54], but activity was low, and substantial growth is typically not seen until water temperature exceeds 20 °C [73]. In the cooler Baltic waters *Nodularia* spp. and *Aphanizomenon* spp. Are observed, which are restricted to 16–22 °C and 25–28 °C, respectively [74]. In warmer waters of the tropics and subtropics, heterocystous diazotrophs (such as *Richelia intracellularis*) are only encountered as symbionts of diatoms like *Hemiaulus* spp., and *Rhizosolenia* spp. [75].

For photoautrophic diazotrophs, nitrogenase activity is intimately linked to photosynthesis [76]. Thus, light is an obvious and important factor potentially regulating or constraining this process. Carpenter [73] has summarized much of the early marine work largely relating to *Trichodesmium* spp. with respect to their relationship to light. Whereas many non-heterocystous cyanobacteria fix nitrogen during the night, and thereby uncouple N_2 fixation from photosynthesis, *Trichodesmium* fixes nitrogen exclusively during the light period and shows a strong diel pattern of activity with maxima during midday [59, 63, 69, 77]. Natural populations of *Trichodesmium*, often found in the upper layers of the euphotic zone, appear to be adapted to high light with a relatively shallow compensation depth (typically 100–200 mmol quanta $m^{-2} s^{-1}$) for photosynthesis [78].

To calculate how much nitrate is produced via N_2-fixation, Redfield stoichiometry can be used [79]. With the given $NO_{3\,exp}^-$ and the measured concentration of the actual dissolved inorganic nitrogen (DIN = $NO_3^- + NO_2^- + NH_4^+$), the amount of NO_3^-, which is removed by denitrification ($NO_{3\,def}^-$), can be calculated as follows:

$$NO_{3\,def}^- = NO_{3\,exp}^- - DIN \tag{11}$$

An alternative stoichiometric method involves the quasi-conservative tracer N* [80, 81]. Gruber and Sarmiento used high quality data from JGOFS and GEOSECS databases to develop a general relationship between fixed inorganic nitrogen (DIN) and phosphate (PO_4^{3-}) for the world's oceans:

$$N^* = DIN - 16 \left[PO_4^{3-} \right] + 2.90 \tag{12}$$

The value of 2.90 is the deviation from the amount of DIN predicted by the Redfield stoichiometry (N:P =16) and the world-ocean N:P regression relationship. Negative values of N* are interpreted to show the net *denitrification* whereas positive values show the net *nitrogen fixation*.

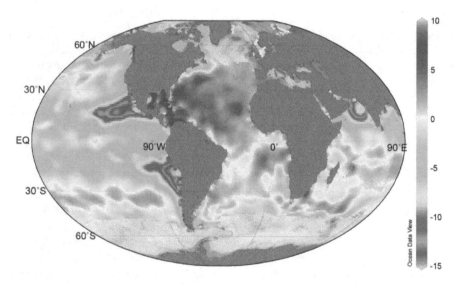

Figure 2. N*= [NO_4^{3-}] − 16*[PO_4^{3-}] + 2.9 on the 26.5 sigma theta surface (World Ocean Atlas 2005). Note the large negative N* values in the Tropical Pacific and Arabian Sea and positive N* values in the Tropical Atlantic Ocean.

Strong positive N* values have been observed in the tropical and subtropical North Atlantic (fig.2) and are proposed to be a result of the input into these areas of diazotrophic (i.e. nitrogen fixer) biomass with a higher N:P content than that typical of eukaryotic phytoplankton of the upper ocean [80-82]. The observation of relatively high concentrations of DON in surface waters of regions of the tropical oceans have also been attributed to nitrogen fixation [83-86]. At the Hawaiian Ocean Time (HOT) series station, pools of DON increased during a period in which microbial nitrogen fixation also became more prominent [87, 88].

Iron limitation may be important for cyanobacterial N_2 fixation as Fe is a cofactor in photosynthesis and nitrogenase, leading to is ~5 times higher Fe requirement for diazotrophy than that for ammonia assimilation [77]. Even though the concentration of dissolved Fe (dFe) is low in many of the environments where N_2 fixation occurs [89-92], a number of studies have speculated on diazotroph response to mineral dust Fe fertilization [93, 94]. A 1999 Saharan dust event coincided with increases in dissolved Fe concentrations on the west Florida shelf and a 100-fold increase in *Trichodesmium* biomass [95]. N_2 fixation rates were not measured, but DON concentrations doubled, presumably due to exudation by N_2 fixers [95]. A Saharan dust addition experiment to surface water samples collected along a west African cruise transect (35°W–17°W), found a minimal increase in CO_2 fixation and a large stimulation in N_2 fixation, suggesting that diazotrophs were co-limited by both P and Fe [96]. It should be noted however, the analysis of aerosol dust shows that while providing Fe, the dust also supplies P and combined N [96-98]. Work on *Trichodesmium* spp. culture has identified mechanisms to deal with P limitation. For instance, under P stress conditions it uptakes a greater amount of DIP and dissolved organic P (DOP), decreasing the requirement for cellular P.

N_2 fixation is an energetically costly process in comparison to ammonia assimilation, thus it is logical to assume that presence of high concentrations of fixed N play a role of an inhibitor. Indeed, under mM concentrations of ammonia (concentrations essentially never observed in the real surface ocean), N_2 fixation is inhibited in most diazotrophs [99]. However, mM concentrations of NO_3^- stopped N_2 fixation in some strains [100] but have little to no effect in others [101]. The higher energy requirement for the nitrate reduction compared to 'direct' assimilation of ammonia might explain these differences. Addition of ammonia to natural samples of *Trichodesmium* spp. can shut down nitrogenase activity [69]. However, pure culture studies have shown mixed results, with some cultures indicating no reduction of their activity [102] and others showing a reduction in activity under short-term additions of fixed N sources [103]. Interestingly, indication for N_2 fixation has been found also in the nitrate-rich upwelling areas off Chile and Peru [104, 105], indicating that N_2 fixation may not be as strongly regulated by fixed N as previously presumed.

4.1.2. Nitrogen fixation in sediments

Within the marine realm, N_2 fixation in sediments has been mostly recognized in pelagic environments or in benthic microbial mats, sea grass sediments, and coral reef systems [106]. A recent study by Bertics et al. [107] revealed that N_2-fixation, coupled to sulfate reduction, is stimulated by bio-irrigation in organic-rich coastal sediments. The authors suggested that N_2-fixation has been largely overlooked in sediments, and that its occurrence might be significant, contradicting the general belief that sediments are areas of net nitrogen loss through denitrification and anaerobic ammonium oxidation [108, 109].

4.2. Nitrification

Nitrification process oxidizes NH_4^+ into NO_2^- and further to NO_3^-. Ammonium rarely occurs at high concentrations in the surface ocean as it is rapidly recycled between heterotrophic and N_2 fixing organisms and many heterotrophic and photosynthetic plankton species. N_2 fixing organisms play a role of ammonia source as they excrete NH_4^+ directly or release organic N that is microbialy degraded to NH_4^+, while plankton utilize ammonia as a N source for building up its biomass. Similarly, accumulation of NO_2^- is rarely observed in oxygenated habitats, although NO_2^- is an essential intermediate in several oxidation and reduction processes in the N cycle.

Nitrification is carried out through a combination of two microbial processes: ammonia oxidation to NO_2^- and nitrite oxidation to NO_3^-. The first step of the process is enabled by nitrifying organisms, ammonium-oxidizing bacteria and archaea (AOB and AOA), while the second step is carried out by nitrite-oxidizing bacteria (NOB). No known organism can carry out both reactions. The overall reaction of NH_4^+ oxidation (eq. 13) for AOB shows that the process consumes molecular oxygen and produces NO_2^- and hydrogen ions. Molecular oxygen is required in the first step of the reaction, which is catalyzed by a monooxygenase (NH_3 monooxygenase, AMO). Oxygen is also consumed by the terminal oxidase as a result of electron transport generating adenosine-5′-triphophate (ATP) for cellular metabolism. The immediate product of AMO is hydroxylamine, which is further oxidized by hydroxylamine oxidoreductase (HAO) to NO_2^-. In contrast, AOA apparently do not possess the hydroxylamine reductase gene, so the pathway of ammonia oxidation in these organisms must be quite different.

$$NH_4^+ + O_2 + H^+ + 2\,e^- \xrightarrow{\text{AMO}} NH_2OH + H_2O \xrightarrow{\text{HAO}} NO_2^- + 5\,H^+ + 4\,e^- \qquad (13)$$

Depending on the conditions, NO, N_2O and even N_2 have been reported as secondary products in autotrophic NH_4^+ oxidation by both marine and terrestrial strains [110, 111]. Although N_2O and NO can be produced in vitro by HAO from hydroxylamine [112], the reduction of NO_2^- appears to be the dominant pathway in all cells [113-115]. One of the first studies of kinetic isotopic effect of nitrification in pure cultures of *Nitrosomonas europaea* was made by the Mariotti group [116], where a fractionation factor of 34.7 ± 2.5‰ was determined for conversion of NH_4^+ into NO_2^-. The later work of Casciotti [117] identified that the fractionation factor for different *Nitrosomonas* cultures of ammonia-oxidizing bacteria was linked to amino acid sequences. The isotopic effect in these cultures ranged from 14.2‰ to 38.2‰.

N_2O is produced by AOB through two separate pathways: hydroxylamine decomposition and nitrite reduction, so-called "nitrifier denitrification" [114]. The isotopic compositions ($\delta^{15}N_{bulk}$, $\delta^{18}O$, $\delta^{15}N_\alpha$, $\delta^{15}N_\beta$, and site preference = $\delta^{15}N_\alpha - \delta^{15}N_\beta$) of the produced N_2O can provide insight into the mechanisms of N_2O production under different conditions [48, 50, 118]. For example, N_2O production through nitrifier denitrification has low $\delta^{15}N_{bulk}$ and low site preferences relative to that produced by hydroxylamine decomposition [119]. As discussed earlier, the N_2O site preference (SP) is determined by the enzymatic mechanism [49, 120]. The site preference of N_2O produced during nitrification is +30‰ to +38‰, while N_2O production via denitrification and nitrifier-denitrification has a SP of -10‰ to +5‰ [48, 118, 121, 122]. The large difference between the SP values of these two mechanisms of N_2O production provides a good tool with which to distinguish their contributions.

The biochemistry of the NO_2^- oxidation is simpler than that of NH_4^+ oxidation because it is only involves a transfer of two electrons without intermediates (eq. 14). The additional oxygen atom in NO_3^- is derived from water, and the molecular oxygen that is involved in the net reaction results from electron transport involving cytochrome oxidase.

$$NO_2^- + H_2O \xrightarrow[oxydoreductase]{nitrite} NO_3^- + 2H^+ + 2e^- + 0.5O_2 \xrightarrow[oxidase]{cytochrome} NO_3^- + H_2O \qquad (14)$$

A recent study of isotopic fractionation during nitrite oxidation to nitrate [1] showed a unique negative isotopic effect of -12.8‰. Because of the inverse isotope effect, when nitrite oxidation is taking place, the $\delta^{15}N_{NO2}$ (and $\delta^{18}O_{NO2}$) values are expected to be lower than the NO_2^- initially produced by ammonia oxidation or nitrate reduction. In most parts of the ocean, however, NO_2^- does not accumulate and the isotope effects of different processes complicate the interpretation. In the euphotic zone NO_2^- is either oxidized to NO_3^- or assimilated into particulate nitrogen (PN) and the heavy isotope can be preferentially shunted in one direction vs. the other [119].

4.2.1. Controlling factors of nitrification and its distribution

Nitrifying bacteria are traditionally considered to be obligate aerobes; they require molecular oxygen for reactions in the N oxidation pathways and for respiration. A significant positive correlation between apparent oxygen utilization (AOU) and N_2O accumulation is often observed in marine systems [123-125]. The relationship implies that nitrification is responsible for N_2O accumulation in oxic waters where it is released as a byproduct. The relationship breaks down at very low oxygen concentrations (~6 mM, [126]), where N_2O is consumed by denitrification.

There is abundant evidence from culture studies that both AOB and NOB are photosensitive. Several studies of nitrification rates in surface seawaters from various geographical regions show profiles that are consistent with light inhibition of both NH_4^+ and NO_2^- oxidation [127-129]. The highest nitrification rates occur in a region near the bottom of the euphotic zone. It is in this interval where nitrifying bacteria can compete with phytoplankton for NH_4^+, as the rates of nutrient assimilation are reduced due to light limitation. A sharp peak of nitrification rate is often observed at water column depth where the light intensity is 5–10% of surface light intensity [118, 127-129].

In both deep and shallow sediments, nitrification can be one of the main sinks for oxygen [130, 131]. In continental shelf sediments, nitrification and denitrification are often closely linked. Coupled nitrification/denitrification is invoked to explain the observation that the rate of N_2 flux out of sediments can greatly exceed the diffusive flux of NO_3^- into the sediments [132]. Ammonium is produced during aerobic and anaerobic remineralization of organic matter. It is then oxidized to NO_3^- and subsequently reduced to N_2. Anaerobic oxidation of NH_4^+ can be a factor supporting the imbalance between supply and consumption of NO_3^-. Nitrification can supply up to 100% of the NO_3^-, which is later consumed by denitrification [133, 134].

Although oxygen and NH_4^+ conditions likely differ between planktonic and sediment environments, there is no clear evidence from clone libraries that water column and sediment nitrifying communities are significantly different in composition and regulation.

To measure nitrification rates, isotope dilution can be used. In this method, the isotope dilution of the NO_2^- pool (in the case of NH_4^+ oxidation) or the NO_3^- pool (in the case of NO_2^- oxidation) is monitored over time [135]. Alternatively, in the ^{15}N method $^{15}NH_4^+$ or $^{15}NO_2^-$ is added to a sample and then the appearance of ^{15}N label in the NO_2^- or NO_3^- pools is measured over time [24]. One disadvantage of the method is that the increase in NH_4^+ or NO_2^- concentrations caused by the tracer addition can result in enhanced NH_4^+ or NO_2^- oxidation rates [129]. When small additions of ^{15}N are used, the isotope dilution of the substrate pool must be monitored over time.

4.3. Nitrogen assimilation

Nitrification and nitrogen assimilation are usually found at similar depths in the ocean as they compete for the same reactants (NH_4^+). At the same time, nitrification, producing NO_2^- and NO_3^-, provides assimilation with additional reactants. Nitrogen assimilation is the process of incorporation of reactive nitrogen species (NO_3^-, NO_2^- and NH_4^+) into the bacterial cell. Ammonium, which is produced in the photic zone by heterotrophic processes, is assimilated immediately by phytoplankton and heterotrophic bacteria before it can be nitrified. Ammonium is often the dominant form of dissolved N assimilated in a variety of marine and estuarine systems. This form of N is energetically efficient for cells to use, because it is already reduced and is a common cellular transient in N metabolism, requiring little additional energy for assimilation. Despite the low NH_4^+ concentrations in oceanic systems, uptake and regeneration of NH_4^+ are tightly coupled and result in rapid turnover times [136].

Nitrate concentrations in the surface ocean are usually maintained at low levels as a) phytoplankton assimilate NO_3^- rapidly and b) nitrate can be supplied by mixing or diffusion from the deep NO_3^- reservoir only at the regions with strong upwelling. The assimilation of NO_3^- is more energetically demanding than NH_4^+, because it requires a synthesis of NO_3^- and NO_2^- reductases, associated active transport systems, and the turnover of cellular ATP and nicotinamide adenine dinucleotide phosphate (NADPH) [136]. In order to assimilate NO_3^- for growth, phytoplankton must first possess the genetic capacity to synthesize the necessary enzymes and transport systems- a capacity which not all phytoplankton have. Further, the supply of NO_3^- is limited by nitrification and vertical mixing.

Since nitrite reductase (NiR) is required for NO_3^- assimilation, organisms that assimilate NO_3^- can by default assimilate NO_2^-. Consequently, some organisms that cannot use NO_3^- can still use NO_2^- (e.g., some clones of *Prochlorococcus*; [137]). It has been suggested that this rather unique N physiology has arisen from the evolutionary loss of genes, which are necessary for the assimilation of NO_3^- [138, 139]. The resultant evolution of many ecotypes specifically adapted for unique environments where NO_2^- production can be high.

Nitrogen isotope fractionation can take place at different stages of the processes. Wada and Hattori [140] first argued that N isotope fractionation by phytoplankton occurs during *nitrate reduction*, while Montoya and McCarthy [141] ultimately favored fractionation associated with *nitrate transport* into the cell. The N isotope data presented in Granger at al. [142] suggest that an isotope effect is driven solely by the reductase, and not by uptake at the cell surface and that this effect results in a fractionation factor of 5 – 10‰. Ammonium assimilation is shown to have higher fractionation factor of +14 – +27‰ [143, 144].

In most parts of the ocean, however, assimilation and nitrification occur concurrently, which additionally complicates the isotope analysis. In general, nitrite oxidation, as discussed above, transfers NO_2^- with an elevated $\delta^{15}N$ to the NO_3^- pool, while nitrite assimilation transfers the residual NO_2^- with a lower $^{15}N{:}^{14}N$ ratio into the PN pool. If the fractionation factor of nitrite assimilation is 1‰ [144] and fractionation factor of nitrite oxidation is -15‰ [1], then the $\delta^{15}N_{NO2}$ at steady state will be lower than the source of NO_2^-, unless nitrite assimilation is >95% of the NO_2^- sink. This has the opposite effect of the ammonia oxidation/assimilation branching where ammonia oxidation transfers low $^{15}N{:}^{14}N$ material into the NO_2^- and NO_3^- pools and higher $^{15}N{:}^{14}N$ material into the PN pool [119].

4.3.1. Factors controlling nitrogen assimilation and its distribution

Most nitrogen assimilation takes place in the surface waters of the ocean. The most important factors controlling assimilation are the presence of oxygen for respiration and/or light for photosynthesis and enzymes responsible for metabolic processes. Most of the enzymes involved in the uptake and assimilation of N are bound to energy sources and thus are affected by light, presence of oxygen, and the supply of enzyme co-factors and metabolic substrates. For example, uptake and reduction of NO_3^-, NO_2^-, and urea have been linked to the light supply in phytoplankton due to the necessity for ATP and NADPH from photo-phosphorylation. While uptake and reduction of these compounds is thought to proceed at maximum rates only in the light under nutrient replete conditions, active uptake of these compounds may occur even in the dark [145].

The accumulation of intracellular product pools, on the other hand, can result in the inhibition of uptake and assimilation. The posttranslational modification of enzymes can regulate uptake and assimilation by modulating the number of active sites available for catalysation of specific reactions. Studies examining the transcriptional activation of enzymes involved in the process have begun to demonstrate clear functional relationships between genomes and ecological capabilities. For example, the presence of NH_4^+ represses protein expression (i.e., enzymes) involved in the assimilation of alternative N sources (e.g., NO_3^- and N_2) in some organisms [146]. Overall, most factors controlling assimilation are directly connected with enzymes, which require substrate compounds or energy and oxygen supply for their metabolic activity.

4.4. Nitrogen loss (denitrification and anammox) in the ocean

The term "nitrogen loss" is commonly used for the processes in which fixed nitrogen (nitrate, nitrite or ammonia) is converted into non-bioavailable N_2 in the absence of oxygen. This includes denitrification process, in which combined nitrogen is reduced to gaseous end products, as well as newly discovered process of anammox (anaerobic ammonium oxidation), which converts nitrite and ammonia into N_2 [147]. Another important process of reduction is DNRA (dissimilatory nitrate reduction to ammonia) will be also discussed in this section as it is an important source of ammonia in the ocean. Even though the reduction of nitrate to ammonia is not a N-loss as such, but it is also taking place in suboxic waters, and the ammonia produced can diffuses up to the anoxic waters and may well be oxidized to N_2 by anammox bacteria [148].

4.4.1. Canonical denitrification

Canonical denitrification is a heterotrophic process in which nitrogen oxides serve as the terminal electron acceptor for organic carbon metabolism [149]. The nitrogen oxides are reduced mainly to molecular nitrogen and some nitrous oxide may be formed as a side product. The sequential reduction is enabled by four well-studied enzyme systems: nitrate reductase, nitrite reductase, nitric oxide reductase and nitrous oxide reductase [150] (eq. 15).

$$NO_3^- \xrightarrow{\text{nitrate reductase}} NO_2^- \xrightarrow{\text{nitrite reductase}} NO \xrightarrow{\text{nitric oxide reductase}} N_2O \xrightarrow{\text{nitrous oxide reductase}} N_2 \qquad (15)$$

The intermediate NO_2^- is known to escape the cell and is frequently found in denitrifying environments; likewise, N_2O also can accumulate externally [151]. This leads to local maxima of both intermediates, which tend to occur near the boundaries of the suboxic zone. Another intermediate, NO, is actually a free radical and, thus, very reactive and highly toxic to most bacteria including denitrifiers [152]. In order to minimize the accumulation of NO within the cell, nitrite reductase and nitric oxide reductase are controlled interdependently at both the transcriptional and enzyme activity levels [152, 153].

4.4.2. Anammox

Anammox bacteria are chemoautotrophic bacteria that fix CO_2 using NO_2^- as the electron donor [154] and are thought to be strict anaerobes. Recent publication on the tolerance of anammox bacteria to oxygen levels, however, showed that they can perform a significant activity even under ~20 μmol/l O_2 concentrations [155]. All anammox species found have evolved a membrane-bound intra-cytoplasmic compartment called the anammoxasome. The anammox-asome membrane is made up of high-density lipids, called ladderanes because of their ladder-like structure, which are thought to be specific to anammox bacteria. The proposed mechanism of anammox involves a hydrazine hydrolase, which catalyzes the combination of hydroxyl ammine and ammonium to form hydrazine. The hydrazine-oxidizing enzyme subsequently oxidized it to N_2 [156, 157]. The ladderane-lipid membrane is thought to act as a barrier to diffusion thus isolating the toxic intermediates of the anammox reaction within the anam-moxasome. Interestingly, for every 1 mol of N_2 produced, 0.3 mol of NO_3^- is also produced alongside from NO_2^-. This reaction is believed to be important for replenishing electrons for the acetyl-CoA carbon fixation pathway [158]. At the same time, these bacteria can perform dissimilatory NO_3^- reduction to NH_4^+ (DNRA), which is subsequently combined with NO_2^- to produce N_2, thus mimicking denitrification [159].

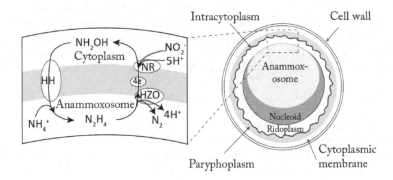

Figure 3. Morphology of the anammox cell and proposed model for the anammox process. HH, hydrazine (N_2H_4) hydrolase; HZO, hydrazine oxidizing enzyme; NR, nitrite reducing enzyme [160].

Anammox bacteria have been identified in many naturally occurring anoxic environments including marine, freshwater, and estuarine sediments [161, 162], oxygen minimum zones [163, 164], soils [165-167], wetlands [168], and anoxic tropical freshwater lakes [169]. These bacteria are not only widespread but may also be very active (e.g. ~67% fixed N-loss [162]). Anammox organisms may live in low abundance but are capable of significant growth and metabolism when a shift in environmental parameters permits anammox communities to grow [170]. The anammox bacteria are slow growing, with a doubling time of ~11 days, even under optimal growth conditions in bioreactors [171]. The state-of-the art method to identify anammox is currently based on isotopic labeling of NH_4^+ or NO_x^-. It relies on the fact that N_2 produced by anammox combines 1:1 atoms from NH_4^+ and NO_2^- [172]. The selective formation of $^{15}N^{14}N$ ($^{29}N_2$) rather than $^{15}N^{15}N$ ($^{30}N_2$) is a clear evidence of anammox activity[162].

4.4.3. Dissimilatory nitrate reduction to ammonium or DNRA

The dissimilatory reduction of nitrate to ammonium is a bacteria-mediated heterotrophic process occurring in anaerobic conditions. Nitrate is first reduced to nitrite and then to NH_4^+. Bock et al. [173] also showed that *N. europaea* and *N. eutropha* were able to nitrify and denitrify at the same time when grown under oxygen limitation ($[O_2]$ ~ 0.2 - 0.4 μmol/l). Under these conditions, oxygen and NO_2^- served simultaneously as electron acceptors and both N_2 and N_2O were produced, whereas under anaerobic conditions N_2 was the predominant end product. This type of nitrification–denitrification pathway may help to explain why ammonium oxidizers remain active in nearly suboxic environments and enhanced N_2O production is observed [174, 175].

The reaction has been reported for sediments underlying anoxic waters [176, 177] and sediments with substantial free sulfide, possibly due to sulfide inhibition of nitrification and denitrification [42]. Suboxic NO_3^- reduction to NH_4^+ appears to occur in environments where there is excess available carbon, such as estuarine sediments [178]. Perhaps more important are the NO_3^- fermenting organisms such as *Thioploca* and *Thiomargarita*, that are found in sediments underlying the major suboxic-denitrifying water columns of the Arabian sea, eastern Tropical Pacific and Namibia [179-181]. These organisms are able to couple the reduction of NO_3^- to NH_4^+ with the oxidation of reduced sulfur compounds. Both *Thioploca* and *Thiomargarita* are able to concentrate NO_3^- at up to 500 mM levels in large vacuoles within their cells for subsequent sulfide oxidation [180, 181].

4.4.4. Controlling factors for N-loss processes and its distribution

The controlling factor for all N-loss processes is molecular oxygen. When the concentration of the denitrification intermediate NO_2^- is plotted against dissolved oxygen, we see that NO_2^- does not appear in the water columns of the Eastern Tropical Pacific until oxygen concentrations are reduced below about 2 μM (see figure 4 [182, 183]).

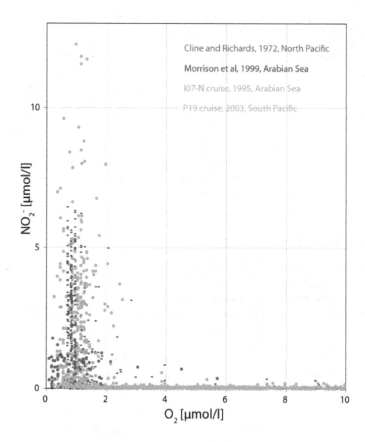

Figure 4. Dissolved oxygen versus nitrite for OMZs. Presented data are taken from publically available cruises I07-N, P19 (http://woce.nodc.noaa.gov/woce_v3/wocedata_2/bathymetry/default.htm) and publications of Cline & Richards [182] and Morisson [183].

Another indication for the denitrification process is consumption of N_2O within the water column, where oxygen concentrations are reduced levels less than c. 5 μM [65]. In oceanic systems the relative importance of heterotrophic denitrification, compared to autotrophic anammox has been debated and believed to depend on the amount of organic matter (OM) available [184, 185]. However, in most of the cases N-loss processes co-exist and it is rather a question which one is playing the first fiddle [186, 187].

About less than half of the current marine denitrification of the global ocean is thought to occur in the three main pelagic Oxygen Minimum Zones (OMZs): Eastern Tropical North Pacific (ETNP), the Eastern Tropical South Pacific (ETSP) and the Northern Arabian Sea (Fig. 4). These zones occur in intermediate waters (~150–1000 m) in locations where the ventilation rate is insufficient to meet the oxygen demand. To calculate nitrogen loss in the water column, the amount of substrate that is consumed in N-loss processes or the amount of N_2 produced should be determined. Similar to estimations of the N-fixation, the Redfield stoichiometry is used to calculate N-deficit or N*. Negative values of N* are interpreted to show the net *N-loss* whereas positive values show the net *nitrogen fixation* (see above).

Devol et al. [188] used N_2:Ar ratios in the Arabian Sea OMZ to determine the amount of N_2 produced during denitrification. The amount of N_2 excess in the OMZ was computed from the increase in the N_2:Ar ratio over that present in the source waters. Measurements of N_2 excess predicted a larger nitrogen anomaly than that estimated by nitrate deficit. The discrepancy is said to be due to incorrect assumptions of the Redfield stoichiometry [189]. Inputs of new nitrogen through N-fixation, N_2 contributions from sedimentary denitrification along continental margins, the anammox reaction, or metal catalyzed denitrification reactions all lead to a shift of the N:P ratio from the Redfield stoichiometry.

Due to N-loss processes, i.e. denitrification and anammox, the $\delta^{15}N$ values in the water column increase to 15‰ or more. For the eastern tropical North Pacific OMZ Brandes et al. [190] calculated the fractionation factors between 22‰ (Arabian Sea closed system model) and 30‰ (eastern tropical North Pacific open-system model). Several authors [190-193] have observed high nitrogen isotopic values and nitrate deficits in the Pacific and Arabian Sea OMZ's and estimated fractionation factor, which all cluster around a value of +25‰. The fractionation factor of N-loss processes in the sediments was shown to be significantly lower (~3 ‰) [194]. Thus, in the waters with strong upwelling, the water column signal can be altered via mixing with water originating from the sediments.

4.4.5. Sedimentary denitrification

Another location where oxygen is typically depleted and denitrification takes place is in marine sediments, especially the continental margin and hemipelagic sediments. Respiratory processes within the sediments act as sinks of oxygen, while overlying water deliver nitrate to the sediment. In most continental shelf sediments, oxygen penetrates to less than 1 cm below the sediment–water interface and even in most deep-sea sediments oxygen penetration is restricted to the approximately upper 10 cm. Thus, ample environments exist for N-loss in marine sediments. Due to the fact that marine sediments are usually rich with organic matter, denitrification and DNRA are the prevailing mechanisms in the sediments [176].

5. Modern ocean nitrogen budget

The discussion of N-budget is usually based on balance or imbalance of nitrogen sources and sinks. Nitrogen sources in this case are N_2-fixation, atmospheric wet and dry deposition, and riverine input. Nitrogen sinks include denitrification and anammox in the water column, sedimentary denitrification, and sedimentation of organic matter. Nitrification and DNRA processes "recycle" nitrogen from ammonia into nitrate and vice versa. Although the primary sources and sinks are relatively well known, the quantification of their fluxes is associated with considerable uncertainty [81, 195-199] (see figure 5). The major problem in calculating global budgets of N sources and sinks is extrapolating measured fluxes from a limited number of oceanic regions to the global ocean.

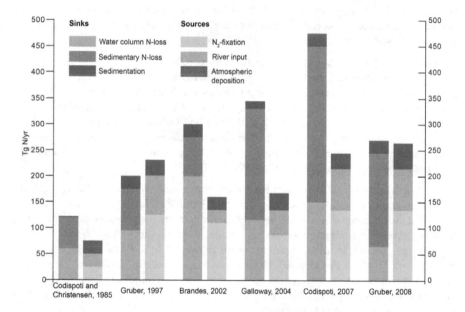

Figure 5. Nitrogen budget estimations in the literature.

Additional complications in calculating N budgets are caused by interplay between and potential coexistence of N cycle processes at the same depth. A schematic water column profile with corresponding processes over O_2, NO_2^-, and N_2O gradients is presented in the figure 6 (based on water column profiles in the South Pacific ocean [65]). In the surface layer, nitrogen assimilation and rapid recycling is taking place, producing nitrite where oxygen and bioavailable N are plentiful enough. Ammonia oxidation (first step of nitrification) lead to efflux of N_2O [200] in the surface waters. Ammonium release and its oxidation in the well-oxygenated waters lead to high production of NO_2^- (primary nitrite maximum). At oxygen concentration ~ 5 µmol/l, denitrification process become more preferential producing to "secondary nitrite maximum" in the OMZ. This NO_2^- can also be a source for anammox process at the top of OMZ, while DNRA process can provide a source of ammonia [148]. N_2O consumption in the water column provides a strong evidence for denitrification producing N_2. At the borders of OMZ nitrification and denitrification processes may coexist [196], leading to production of N_2O at these water depths.

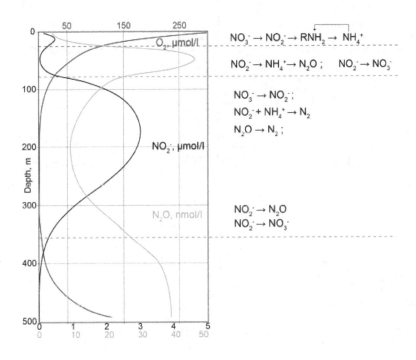

Figure 6. Schematic water column profile with corresponding processes over O_2, NO_2^-, and N_2O gradients.

Nitrogen isotope ratios in the water column and in sedimentary records provide additional constraints for nitrogen budget calculation. Average $\delta^{15}N$ of fixed nitrogen in the deep ocean is close to 5‰. The sedimentary records show that this global average increased during deglaciation periods, when stronger denitrification was taking place. However, the global average $\delta^{15}N$ range for the last 30.000 years was in the range 4-6‰ [201], which suggests that N sources and sinks should be balanced, at least the millennial time scale. In this case, the processes increasing $\delta^{15}N$ (assimilation, N-loss) has to be balanced with those, which are decreasing it (N_2-fixation).

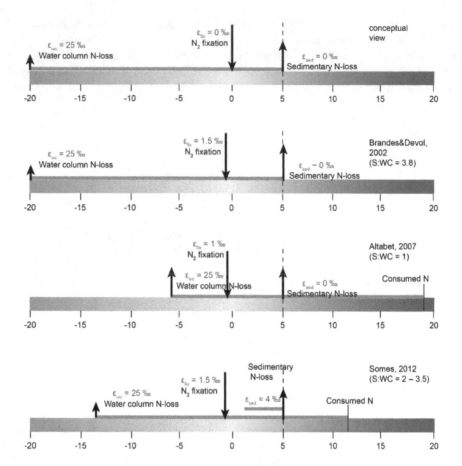

Figure 7. Nitrogen isotope range and processes influencing on the ^{15}N budget. Dashed line at 5‰ represent an oceanic average value [197, 202, 203].

The conceptual model of the nitrogen isotope budget is initially based on two processes: N_2-fixation and denitrification. Sedimentary denitrification was believed not to change the isotopic signature (ε_{sed}) of nitrogen due to N-limitation in the sediments, while water column denitrification (ε_{wc}) has a fractionation factor of ε between 25 and 35 ‰. The N_2- fixation fractionation factor ε_{fix} is close to 0‰ as it incorporates nitrogen from the atmosphere. An early attempt to construct a one-dimensional model was undertaken by Brandes and Devol [197], where they set fractionation factor to ε_{fix} and ε_{sed} to 1.5‰; and ε_{wc} to 25‰. On the figure 8 scale of nitrogen isotope is illustrated. In the Brandes and Devol model, sedimentary N-loss does not change the ^{15}N budget substantially (ε_{sed} in the range of 0 to 1.5‰). Thus, a large amount of sedimentary denitrification is necessary to balance water column denitrification and bring overall isotopic values of the losses into the range of sources, leading to the value of Sediment to Water column denitrification (W:WC) ratio of 3.8. The authors noticed that with fluxes estimations of total N-loss of 400 Tg N/yr and fixation of 100 Tg N/yr, "the combined budget would be out of balance by over -200 Tg N/yr" [197]. This implies that fixation rates should be 2–3 times higher than 100 Tg N/yr, which was an accepted value at that time. Recent observations that N_2- fixation is indeed underestimated by up to two fold support these calculations.

More recent calculation of the nitrogen isotope budget in year 2007 included so-called "consumed-N" into the model, which represents nitrogen utilization in the euphotic zone [202]. During N utilization, nitrogen fractionation takes place, shifting the "initial" value of the ambient NO_3^- from +5‰ to 17 ‰. Water column N-loss leads to production of N_2 gas with $\delta^{15}N$ values 25‰ lighter than the "initial". This process shifts the nitrogen isotope budget significantly and even leads to balance of water column and sedimentary denitrification (S:CW =1). A recent model of the pre-industrial N-budget includes a sedimentary fractionation factor ε_{sed} of 4‰ and ε_{fix} of 1.5‰ [203]. The value of "consumed-N" was set to +12‰, taking into account the fractionation factor of utilization of 7‰. In this model, the S:CW denitrification ratio varies from 2 to 3.5. Current budget estimates can, however, differ from these prior estimates due to an increase of riverine and atmospheric input of nitrogen into the ocean [198].

6. Conclusions

Of the two approaches whereby nitrogen isotope measurements are used to study marine N cycle processes, stable isotope natural abundance measurements have the great advantage of providing information on integrated processes with little or no manipulation of the system. In contrast tracer-based rate measurements are better used for rate measurements. The natural abundance and tracer approaches are strongly complementary, providing insight into the nitrogen cycle on very different spatial and temporal scales. The magnitude of processes that have distinctive isotopic fractionation patterns can potentially be quantified if the fractionations are known accurately. Thus, there is a significant ongoing need for careful measurement of isotopic fractionation patterns under controlled laboratory conditions as well as under field conditions, which in the best case should be combined. In order to interpret isotopic distributions and trends and to construct meaningful isotopic budgets, the openness of the system (ventilation rates) and the nature of N-exchanges should be known.

The growing set of $\delta^{15}N$ measurements makes it increasingly feasible to incorporate isotopic constraints into ecosystem and biogeochemical models of nitrogen in the ocean. A robust isotope budget for the ocean will require extension of these sampling efforts. Time-series measurements from a variety of locations will also greatly enhance the utility of nitrogen isotope measurements, particularly in quantifying the flux of nitrogen through the biota during blooms and other transient events. Such transient events are effectively natural tracer experiments created by isotopic fractionation during phytoplankton consumption of inorganic nitrogen and could provide an ecosystem-level analog to the short-term tracer experiments typically carried out in small volumes aboard ship. Complementing nitrogen stable isotope measurements with focused rate measurements may ultimately provide the best strategy for studying the dynamics of the nitrogen cycle across a variety of temporal and spatial scales.

Author details

Evgenia Ryabenko[1,2*]

1 GEOMAR/Helmholtz Centre for Ocean Research Kiel, Kiel, Germany

2 Institute of Groundwater Ecology, Helmholtz Zentrum München – Germany Research Center for Environmental Health, München, Germany

References

[1] Casciotti, K. L. *Inverse kinetic isotope fractionation during bacterial nitrite oxidation.* Geochimica et Cosmochimica Acta, (2009). , 2061-2076.

[2] Nier, A. O. *A redetermination of the relative abundances of the isotopes of carbon, nitrogen, oxygen, argon, and potassium.* Physical Review, (1950). , 789-793.

[3] Coplen, T. B, Krouse, H. R, & Böhlke, J. K. *Reporting of nitrogen isotope abundances- (Technical report).* Pure and Applied Chemistry, (1992). , 907-908.

[4] Hoefs, J. *Stable Isotope Geochemistry* 6ed(2009). Berlin: Springer Verlag.

[5] Owens, N. J. P, & Rees, A. P. *Determination of nitrogen-15 at sub-microgram levels of nitrogen using automated continuous flow isotope ratio mass spectrometry.* Analyst, (1989). , 1655-1657.

[6] Ryabenko, E, Altabet, M. A, & Wallace, D. W. R. *Effect of chloride on the chemical conversion of nitrate to nitrous oxide for δ^{15}N analysis.* Limnology and Oceanography-Methods, (2009). , 545-552.

[7] Mcilvin, M. R, & Altabet, M. A. *Chemical conversion of nitrate and nitrite to nitrous oxide for nitrogen and oxygen isotopic analysis in freshwater and seawater.* Analytical Chemistry, (2005). , 5589-5595.

[8] Sigman, D. M, et al. *A bacterial method for the nitrogen isotopic analysis of nitrate in seawater and freshwater.* Analytical Chemistry (2001). , 4145-4153.

[9] Casciotti, K. L, & Mcilvin, M. R. *Isotopic analyses of nitrate and nitrite from reference mixtures and application to Eastern Tropical North Pacific waters.* Marine Chemistry, (2007). , 184-201.

[10] Casciotti, K. L, et al. *Measurement of the oxygen isotopic composition of nitrate in seawater and freshwater using the denitrifier method.* Analytical Chemistry, (2002). , 4905-4912.

[11] Granger, J, et al. *A method for nitrite removal in nitrate N and O isotope analyses.* Limnology and Oceanogrophy-Methods, (2006). July): , 205-212.

[12] Glibert, P. M, et al. *Isotope-dilution models of uptake and remineralization of ammonium by marine plankton.* Limnology and Oceanography, (1982)., 639-650.

[13] Harrison, W. G, & Harris, L. R. *Isotope-dilution and its effects on measurements of nitrogen and phosphorus uptake by oceanic microplankton.* Marine Ecology-Progress Series, (1986)., 253-261.

[14] Mohr, W, et al. *Methodological underestimation of oceanic nitrogen fixation rates.* Plos One, (2010)., e12583.

[15] Frew, N. *The role of organic films in air-sea gas exchange,* in *The sea surface and globale change,* R.A. Duce and P.S. Liss, Editors. (1997). Cambridge University Press: Cambridge, UK., 121-172.

[16] Asher, W. E, & Pankow, J. F. *Prediction of gas/water mass transport coefficients by a surface renewal model* Environmental Science & Technology, (1991)., 1294-1300.

[17] Mariotti, A. *Atmospheric nitrogen is a reliable standard for natural ^{15}N abundance measurements.* Nature, (1983)., 685-687.

[18] Montoya, J. P.: Nitrogen stable isotopes in marine environments, in: Nitrogen in Marine Environment, edited by: Capone, D. G., Bronk, D. A., Mulholland, M. R., and Carpenter, E. J., Elsevier Inc, Amsterdam, 1277-1302, 2008.

[19] Wilkerson, F. P, & Grunseich, G. *Formation of blooms by the symbiotic ciliate Mesodinium rubrum- The significance of nitrogen uptake* Journal of Plankton Research, (1990)., 973-989.

[20] Raimbault, P, Slawyk, G, & Garcia, N. *Comparison between chemical and isotopic measurements of biological nitrate utilization: further evidence of low new-production levels in the equatorial Pacific.* Marine Biology, (2000)., 1147-1155.

[21] Libby, S. P, & Wheeler, P. A. *Particulate and dissolved organic nitrogen in the central and eastern equatorial Pacific.* Deep Sea Research Part I: Oceanographic Research Papers, (1997)., 345-361.

[22] Taguchi, S, & Laws, E. A. *On the microparticles which pass through glass-fiber filter type GF/F in coastal and open waters.* Journal of Plankton Research, (1988)., 999-1008.

[23] Maske, H, & Garcia-mendoza, E. *Adsorption of dissolved organic matter to the inorganic filter substrate and its implications for ^{14}C uptake measurements.* Applied and Environmental Microbiology, (1994)., 3887-3889.

[24] Raimbault, P, & Slawyk, G. *A Semiautomatic wet-oxidation method for the determination of particlulate organic nitrogen collected on filters.* Limnology and Oceanography, (1991)., 405-408.

[25] Olson, R. J. *Differential photoinhibition of marine nitrifying bacteria: A possible mechanism for the formation of the primary nitrite maximum.* Journal of Marine Research, (1981)., 227-238.

[26]   Yakushiji, H, & Kanda, J. *Determination of experimentally enriched ^{15}N in nitrate nitrogen based on an improved method of azo dye formation.* Journal of Oceanography, (1998). , 337-342.

[27]   Bronk, D. A, & Ward, B. B. *Gross and net nitrogen uptake and DON release in the euphotic zone of Monterey Bay, California.* Limnology and Oceanography, (1999). , 573-585.

[28]   Sigman, D. M, et al. *Natural abundance-level measurement of the nitrogen isotopic composition of oceanic nitrate: an adaptation of the ammonia diffusion method.* Marine Chemistry, (1997). , 227-242.

[29]   Tanaka, T, & Saino, T. *Modified method for the analysis of nitrogen isotopic composition of oceanic nitrate at low concentration.* Journal of Oceanography, (2002). , 539-546.

[30]   Risgaard-petersen, N, et al. *Application of the isotope pairing technique in sediments where anammox and denitrification coexist.* Limnology and Oceanography-Methods, (2003). , 63-73.

[31]   Jensen, K. M, Jensen, M. H, & Cox, R. P. *Membrane inlet mass spectrometric analysis of N-isotope labelling for aquatic denitrification studies.* Fems Microbiology Ecology, (1996). , 101-109.

[32]   Feast, N. A, & Dennis, P. F. *A comparison of methods for nitrogen isotope analysis of groundwater.* Chemical Geology, (1996). , 167-171.

[33]   Preston, T, et al. *Isotope dilution analysis of combined nitrogen in natural waters.1. Ammonium.* Rapid Communications in Mass Spectrometry, (1996). , 959-964.

[34]   Lehmann, M. F, Bernasconi, S. M, & Mckenzie, J. A. *A method for the extraction of ammonium from freshwaters for nitrogen isotope analysis.* Analytical Chemistry, (2001). , 4717-4721.

[35]   Deen, O, & Porter, W. A. a. n. d L. K. *Devarda's alloy reduction of nitrate and tube diffusion of the reduced nitrogen for indophenol ammonium and nitrogen-15 determinations.* Analytical Chemistry, (1980). , 1164-1166.

[36]   Holmes, R. M, et al. *Measuring ^{15}N-NH_4^+ in marine, estuarine and fresh waters: An adaptation of the ammonia diffusion method for samples with low ammonium concentrations.* Marine Chemistry, (1998). , 235-243.

[37]   Risgaard-petersen, N, Revsbech, N. P, & Rysgaard, S. *Combined microdiffusion-hypobromite oxidation method for determining nitrogen-15 isotope in ammonium.* Soil Science Society of America Journal (1995). , 1077-1080.

[38]   Sebilo, M, et al. *The use of the 'Ammonium diffusion' method for $\delta^{15}N$-NH_4^+ and $\delta^{15}N$-3-measurementsComparison with other techniques.* Environmental Chemistry, (2004). , 99-103.

[39] Dudek, N, Brzezinski, M. A, & Wheeler, P. A. *Recovery of ammonium nitrogen by solvent extraction for the determination of relative ^{15}N abundance in regeneration experiments.* Marine Chemistry, (1986). , 59-69.

[40] Zhang, L, et al. *Sensitive measurement of NH_4^{+} $^{15}N/^{14}N$ ($\delta^{15}NH_4^{+}$) at natural abundance levels in fresh and saltwaters.* Analytical Chemistry, (2007).

[41] Steingruber, S. M, et al. *Measurement of denitrification in sediments with the N-15 isotope pairing technique.* Applied and Environmental Microbiology, (2001). , 3771-3778.

[42] Capone, D. G, & Montoya, J. P. *Nitrogen fixation and denitrification,* in *Methods in Marine Microbiology,* J. Paul, Editor (2001). Academic Press: New York. , 501-515.

[43] An, S. M, Gardner, W. S, & Kana, T. *Simultaneous measurement of denitrification and nitrogen fixation using isotope pairing with membrane inlet mass spectrometry analysis.* Applied and Environmental Microbiology, (2001). , 1171-1178.

[44] Kana, T. M, et al. *Denitrification in estuarine sediments determined by membrane inlet mass spectrometry.* Limnology and Oceanography, (1998). , 334-339.

[45] Hartnett, H. E, & Seitzinger, S. P. *High-resolution nitrogen gas profiles in sediment pore-waters using a new membrane probe for membrane-inlet mass spectrometry.* Marine Chemistry, (2003). , 23-30.

[46] Kana, T. M, et al. *Membrane Inlet Mass Spectrometer for rapid high-precision determination of N_2, O_2, and Ar in environmental water samples.* Analytical Chemistry, (1994). , 4166-4170.

[47] Punshon, S, & Moore, R. M. *Nitrous oxide production and consumption in a eutrophic coastal embayment.* Marine Chemistry, (2004). , 37-51.

[48] Casciotti, K. L, McIlvin, M, & Buchwald, C. *Oxygen isotopic exchange and fractionation during bacterial ammonia oxidation* Limnology and Oceanography, (2010). , 753-762.

[49] Frame, C. H, & Casciotti, K. L. *Biogeochemical controls and isotopic signatures of nitrous oxide production by a marine ammonia-oxidizing bacterium.* Biogeosciences (2010). , 3019-3059.

[50] Toyoda, S, & Yoshida, N. *Determination of nitrogen isotopomers of nitrous oxide on a modified isotope ratio mass spectrometer.* Analytical Chemistry, (1999). , 4711-4718.

[51] Sutka, R. L, et al. *Distinguishing nitrous oxide production from nitrification and denitrification on the basis of isotopomer abundances.* Applied and Environmental Microbiology, (2006). , 638-644.

[52] Ostrom, N, & Ostrom, P. *The isotopomers of nitrous oxide: Analytical considerations and application to resolution of microbial production pathways,* in *Handbook of Environmental Isotope Geochemistry,* M. Baskaran, Editor (2011). Springer Berlin Heidelberg. , 453-476.

[53] Ward, B. B, & Bronk, D. A. *Net nitrogen uptake and DON release in surface waters: impor-
      tance of trophic interactions implied from size fractionation experiments.* Marine Ecology-
      Progress Series, (2001). , 11-24.

[54] Brandes, J. A, Devol, A. H, & Deutsch, C. *New developments in the marine nitrogen cy-
      cle.* Chemical Reviews, (2007). , 577-589.

[55] Mccarthy, J. J, & Carpenter, E. J. *Oscillatoria (Trichodesmium) thiebautii (Cyanophyta) in
      the central north Atlantic Ocean.* Journal of Phycology, (1979). , 75-82.

[56] Seitzinger, S. P, Sanders, R. W, & Styles, R. *Bioavailability of DON from natural and an-
      thropogenic sources to estuarine plankton.* Limnology and Oceanography, (2002). ,
      353-366.

[57] Montoya, J. P, Carpenter, E. J, & Capone, D. G. *Nitrogen fixation and nitrogen isotope
      abundances in zooplankton of the oligotrophic North Atlantic.* Limnology and Oceanogra-
      phy, (2002). , 1617-1628.

[58] Zehr, J, & Paerl, H. *Biological nitrogen fixation in the marine environment.,* in *Microbial
      Ecology of the Oceans.,* D.L. Kirchman, Editor (2008). Wiley-Liss, Inc.: New York. ,
      481-525.

[59] Carpenter, E. J. *Nitrogen fixation by Oscillatoria (Trichodesmium) thiebautii in the south-
      western Sargasso Sea.* Deep-Sea Research, (1973). , 285-288.

[60] Saino, T, & Hattori, A. *Diel variation in nitrogen fixation by a marine blue-green alga, Tri-
      chodesmium thiebautii.* Deep-Sea Research, (1978). , 1259-1263.

[61] Carpenter, E. J. *Nitrogen fixation by a blue-green epiphyte on pelagic Sargassum.* Science,
      (1972). , 1207-1209.

[62] Mague, T. H, Mague, F. C, & Holmhansen, O. *Physiology and chemical composition of
      nitrogen-fixing phytoplankton in the central north Pacific ocean.* Marine Biology, (1977). ,
      213-227.

[63] Villareal, T. A. *Widespread occurrence of the Hemiaulus-cyanobacterial symbiosis in the
      southwest north Atlantic ocean.* Bulletin of Marine Science, (1994). , 1-7.

[64] Berman-frank, I, et al. *Segregation of nitrogen fixation and oxygenic photosynthesis in the
      marine cyanobacterium Trichodesmium.* Science, (2001). , 1534-1537.

[65] Wada, E, & Hattori, A. *Natural abundance of ^{15}N in particulate organic matter in the
      North Pacific Ocean.* Geochimica et Cosmochimica Acta, (1976). , 249-251.

[66] Ryabenko, E, et al. *Contrasting biogeochemistry of nitrogen in the Atlantic and Pacific oxy-
      gen minimum zones.* Biogeosciences (2012). , 203-215.

[67] Lehmann, M. F, et al. *Origin of the deep Bering Sea nitrate deficit: Constraints from the
      nitrogen and oxygen isotopic composition of water column nitrate and benthic nitrate fluxes.*
      Global Biogeochemical Cycles, (2005). , GB4005.

[68] Delwiche, C. C, et al. *Nitrogen isotope distribution as a presumptive indicator of nitrogen-fixation.* Botanical Gazette, (1979). , S65-S69.

[69] Zehr, J. P, Carpenter, E. J, & Villareal, T. A. *New perspectives on nitrogen-fixing microorganisms in tropical and subtropical oceans.* Trends in Microbiology, (2000). , 68-73.

[70] Capone, D. G, et al. *Basis for diel variation in nitrogenase activity in the marine planktonic cyanobacterium Trichodesmium thiebautii.* Applied and Environmental Microbiology, (1990). , 3532-3536.

[71] Howarth, R. W, Marino, R, & Cole, J. J. *Nitrogen-fixation in fresh-water, estuarine, and marine ecosystems.2. Biogeochemical controls.* Limnology and Oceanography, (1988). , 688-701.

[72] Howarth, R. W, et al. *Nitrogen fixation in freshwater, estuarine, and marine ecoystems. 1. rates and importance.* Limnology and Oceanography, (1988). , 469-687.

[73] Karl, D, et al. *Dinitrogen fixation in the world's oceans.* Biogeochemistry, (2002). , 47-98.

[74] Carpenter, E. J. *Physiology and ecology of marine Oscillatoria (Trichodesmium).* Marine Biology Letters, (1983). , 69-85.

[75] Lehtimaki, J, et al. *Growth, nitrogen fixation, and nodularin production by two Baltic sea cyanobacteria.* Applied and Environmental Microbiology, (1997). , 1647-1656.

[76] Sohm, J. A, Webb, E. A, & Capone, D. G. *Emerging patterns of marine nitrogen fixation.* Nature Reviews Microbiology, (2011). , 499-508.

[77] Gallon, J. R. *N_2 fixation in phototrophs: adaptation to a specialized way of life.* Plant and Soil, (2001). , 39-48.

[78] Berman-frank, I, et al. *Iron availability, cellular iron quotas, and nitrogen fixation in Trichodesmium.* Limnology and Oceanography, (2001). , 1249-1260.

[79] LaRocheJ. and E. Breitbarth, *Importance of the diazotrophs as a source of new nitrogen in the ocean.* Journal of Sea Research, (2005). , 67-91.

[80] Codispoti, L. A, & Richards, F. A. *An analysis of the horizontal regime of denitrification in the eastern tropical North Pacific.* Limnology and Oceanography, (1976). , 379-388.

[81] Deutsch, C, et al. *Denitrification and N_2 fixation in the Pacific Ocean.* Global Biogeochemical Cycles, (2001). , 483-506.

[82] Gruber, N, & Sarmiento, J. L. *Global patterns of marine nitrogen fixation and denitrification.* Global Biogeochemical Cycles, (1997). , 235-266.

[83] Michaels, A. F, et al. *Inputs, losses and transformations of nitrogen and phosphorus in the pelagic North Atlantic Ocean.* Biogeochemistry, (1996). , 181-226.

[84] Hansell, D. A, & Feely, R. A. *Atmospheric intertropical convergence impacts surface ocean carbon and nitrogen biogeochemistry in the western tropical Pacific.* Geophysical Research Letters, (2000). , 1013-1016.

[85] Hansell, D. A, & Waterhouse, T. Y. *Controls on the distributions of organic carbon and nitrogen in the eastern Pacific Ocean.* Deep-Sea Research Part I-Oceanographic Research Papers, (1997). , 843-857.

[86] Vidal, M, Duarte, C. M, & Agusti, S. *Dissolved organic nitrogen and phosphorus pools and fluxes in the central Atlantic Ocean.* Limnology and Oceanography, (1999). , 106-115.

[87] Abell, J, Emerson, S, & Renaud, P. *Distributions of TOP, TON and TOC in the North Pacific subtropical gyre: Implications for nutrient supply in the surface ocean and remineralization in the upper thermocline.* Journal of Marine Research, (2000). , 203-222.

[88] Karl, D, et al. *The role of nitrogen fixation in biogeochemical cycling in the subtropical North Pacific Ocean.* Nature, (1997). , 533-538.

[89] Karl, D. M, et al. *Ecosystem changes in the North Pacific subtropical gyre attributed to the 1991-92 El-Nino.* Nature, (1995). , 230-234.

[90] Bergquist, B. A, & Boyle, E. A. *Dissolved iron in the tropical and subtropical Atlantic Ocean.* Global Biogeochemical Cycles, (2006).

[91] Blain, S, Bonnet, S, & Guieu, C. *Dissolved iron distribution in the tropical and sub tropical South Eastern Pacific.* Biogeosciences, (2008). , 269-280.

[92] Boyle, E. A, et al. *Iron, manganese, and lead at Hawaii Ocean Time-series station ALOHA: Temporal variability and an intermediate water hydrothermal plume.* Geochimica et Cosmochimica Acta, (2005). , 933-952.

[93] Wu, J, et al. *Soluble and colloidal iron in the oligotrophic North Atlantic and North Pacific.* Science, (2001). , 847-849.

[94] Mahaffey, C, et al. *Biogeochemical signatures of nitrogen fixation in the eastern North Atlantic.* Geophysical Research Letters, (2003). , 1300.

[95] Johnson, K. S, et al. *Surface ocean-lower atmosphere interactions in the Northeast Pacific Ocean Gyre: Aerosols, iron, and the ecosystem response.* Global Biogeochemical Cycles, (2003). , 1063.

[96] Lenes, J. M, et al. *Iron fertilization and the Trichodesmium response on the West Florida shelf.* Limnology and Oceanography, (2001). , 1261-1277.

[97] Mills, M. M, et al. *Iron and phosphorus co-limit nitrogen fixation in the eastern tropical North Atlantic.* Nature, (2004). , 292-294.

[98] Baker, A. R, et al. *Atmospheric deposition of nutrients to the Atlantic Ocean.* Geophysical Research Letters, (2003). , 2296-2300.

[99] Ridame, C, & Guieu, C. *Saharan input of phosphate to the oligotrophic water of the open western Mediterranean Sea.* Limnology and Oceanography, (2002)., 856-869.

[100] Ramos, J. L, Madueno, F, & Guerrero, M. G. *Regulation of nitrogenase levels in Anabaena sp. ATCC 33047 and other filamentous cyanobacteria* Archives of Microbiology, (1985)., 105-111.

[101] Martín-nieto, J, Herrero, A, & Flores, E. *Control of nitrogenase mRNA levels by products of nitrate assimilation in the cyanobacterium Anabaena sp. Strain PCC 7120.* Plant Physiology, (1991)., 825-828.

[102] Sanzalferez, S, & Delcampo, F. F. *Relationship between nitrogen fixation and nitrate metabolism in the Nodularia strains M1 and M2.* Planta, (1994)., 339-345.

[103] Mulholland, M. R, Ohki, K, & Capone, D. G. *Nutrient controls on nitrogen uptake and metabolism by natural populations and cultures of Trichodesmium (Cyanobacteria).* Journal of Phycology, (2001)., 1001-1009.

[104] Holl, C. M, & Montoya, J. P. *Interactions between nitrate uptake and nitrogen fixation in continuous cultures of the marine diazotroph Trichodesmium (Cyanobacteria).* Journal of Phycology, (2005)., 1178-1183.

[105] Raimbault, P, & Garcia, N. *Evidence for efficient regenerated production and dinitrogen fixation in nitrogen-deficient waters of the South Pacific Ocean: impact on new and export production estimates.* Biogeosciences, (2008)., 323-338.

[106] Fernandez, C, Farias, L, & Ulloa, O. *Nitrogen fixation in denitrified marine waters.* Plos One, (2011)., e20539.

[107] Carpenter, E. J, & Capone, D. G. *Nitrogen fixation in the marine environment,* in *Nitrogen in the marine environment,* D.G. Capone, et al., Editors. (2008). Elsevier., 141-198.

[108] Bertics, V. J, & Ziebis, W. *Bioturbation and the role of microniches for sulfate reduction in coastal marine sediments.* Environmental Microbiology, (2010)., 3022-3034.

[109] Gilbert, F, et al. *Hydrocarbon influence on denitrification in bioturbated Mediterranean coastal sediments.* Hydrobiologia, (1997)., 67-77.

[110] Herbert, R. A. *Nitrogen cycling in coastal marine ecosystems.* FEMS Microbiology Reviews, (1999)., 563-590.

[111] Schmidt, I, Van Spanning, R. J. M, & Jetten, M. S. M. *Denitrification and ammonia oxidation by Nitrosomonas europaea wild-type, and NirK- and NorB-deficient mutants.* Microbiology-Sgm, (2004)., 4107-4114.

[112] Zart, D, & Bock, E. *High rate of aerobic nitrification and denitrification by Nitrosomonas eutropha grown in a fermentor with complete biomass retention in the presence of gaseous 2or NO.* Archives of Microbiology, (1998)., 282-286.

[113] Hooper, A. B, & Terry, K. R. *Hydroxylamine oxidoreductase of Nitrosomonas: Production of nitric oxide from hydroxylamine.* Biochimica et Biophysica Acta (BBA)- Enzymology, (1979)., 12-20.

[114] Hooper, A. B, et al. *Enzymology of the oxidation of ammonia to nitrite by bacteria.* Antonie Van Leeuwenhoek International Journal of General and Molecular Microbiology, (1997)., 59-67.

[115] Poth, M, & Focht, D. D. *N kinetic analysis of N$_2$O production by Nitrosomonas europaea: an examination of nitrifier denitrification.* Applied and Environmental Microbiology, (1985)., 1134-1141.

[116] Remde, A, & Conrad, R. *Production of nitric oxide in Nitrosomonas europaea by reduction of nitrite.* Archives of Microbiology, (1990)., 187-191.

[117] Mariotti, A, et al. *Experimental determination of nitrogen kinetic isotope fractionation- Some principles- Illustration for the denitrification and nitrification processes.* Plant and Soil, (1981)., 413-430.

[118] Casciotti, K. L, Sigman, D. M, & Ward, B. B. *Linking diversity and stable isotope fractionation in ammonia-oxidizing bacteria.* Geomicrobiology Journal, (2003)., 335-353.

[119] Sutka, R. L, et al. *Stable nitrogen isotope dynamics of dissolved nitrate in a transect from the North Pacific Subtropical Gyre to the Eastern Tropical North Pacific.* Geochimica et Cosmochimica Acta, (2004)., 517-527.

[120] Casciotti, K. L, & Buchwald, C. *Insights on the marine microbial nitrogen cycle from isotopic approaches to nitrification.* Frontiers in Microbiology, (2012).

[121] Schmidt, H. L, et al. *Is the isotopic composition of nitrous oxide an indicator for its origin from nitrification or denitrification? A theoretical approach from referred data and microbiological and enzyme kinetic aspects.* Rapid Communications in Mass Spectrometry, (2004)., 2036-2040.

[122] Toyoda, S, et al. *Fractionation of N$_2$O isotopomers during production by denitrifier.* Soil Biology & Biochemistry, (2005)., 1535-1545.

[123] Sutka, R. L, et al. *Nitrogen isotopomer site preference of N$_2$O produced by Nitrosomonas europaea and Methylococcus capsulatus Bath.* Rapid Communications in Mass Spectrometry, (2003)., 738-745.

[124] Nevison, C. D, et al. *Interannual and seasonal variability in atmospheric N$_2$O.* Global Biogeochemical Cycles, (2007)., GB3017.

[125] Cohen, Y, & Gordon, L. I. *Nitrous oxide in the oxygen minimum of the eastern tropical North Pacific: evidence for its consumption during denitrification and possible mechanisms for its production.* Deep Sea Research, (1978)., 509-524.

[126] Yoshinari, T. *Nitrous oxide in the sea.* Marine Chemistry, (1976)., 189-202.

[127] Nevison, C, Butler, J. H, & Elkins, J. W. *Global distribution of N₂O and the delta N₂O-AOU yield in the subsurface ocean.* Global Biogeochemical Cycles, (2003). , 1119.

[128] Lipschultz, F, et al. *Bacterial transformations of inorganic nitrogen in the oxygen-deficient waters of the Eastern Tropical South Pacific Ocean.* Deep Sea Research Part A. Oceanographic Research Papers, (1990). , 1513-1541.

[129] Ward, B. B. *Nitrogen transformations in the Southern California Bight.* Deep-Sea Research Part A-Oceanographic Research Papers, (1987). , 785-805.

[130] Ward, B. B, Talbot, M. C, & Perry, M. J. *Contributions of phytoplankton and nitrifying bacteria to ammonium and nitrite dynamics in coastal waters.* Continental Shelf Research, (1984). , 383-398.

[131] Blackburn, T. H, & Blackburn, N. D. *Coupling of cycles and global significance of sediment diagenesis* Marine Geology, (1993). , 101-110.

[132] Grundmanis, V, & Murray, J. W. *Nitrification and denitrification in marine sediments from Puget Sound.* Limnology and Oceanography, (1977). , 804-813.

[133] Devol, A. H, & Christensen, J. P. *Benthic fluxes and nitrogen cycling in sediments of the continental margin of the eastern North Pacific* Journal of Marine Research, (1993). , 345-372.

[134] Laursen, A. E, & Seitzinger, S. P. *The role of denitrification in nitrogen removal and carbon mineralization in Mid-Atlantic Bight sediments.* Continental Shelf Research, (2002). , 1397-1416.

[135] Lehmann, M. F, Sigman, D. M, & Berelson, W. M. *Coupling the ¹⁵N/¹⁴N and ¹⁸O/¹⁶O of nitrate as a constraint on benthic nitrogen cycling.* Marine Chemistry, (2004). , 1-20.

[136] Clark, D. R, Rees, A. P, & Joint, I. *A method for the determination of nitrification rates in oligotrophic marine seawater by gas chromatography/mass spectrometry.* Marine Chemistry, (2007). , 84-96.

[137] Mulholland, M. R, & Lomas, M. W. *Nitogen uptake and assimilation,* in *Nitrogen in the Marine Environment,* D.G. Capone, et al., Editors. (2008). Elsevier Inc.: Amsterdam et al. , 303-384.

[138] Moore, J. K, et al. *An intermediate complexity marine ecosystem model for the global domain.* Deep-Sea Research Part II-Topical Studies in Oceanography, (2002). , 403-462.

[139] Garcia-fernandez, J. M, De Marsac, N. T, & Diez, J. *Streamlined regulation and gene loss as adaptive mechanisms in Prochlorococcus for optimized nitrogen utilization in oligotrophic environments.* Microbiology and Molecular Biology Reviews, (2004). , 630-638.

[140] Garcia-fernandez, J. M, & Diez, J. *Adaptive mechanisms of nitrogen and carbon assimilatory pathways in the marine cyanobacteria Prochlorococcus.* Research in Microbiology, (2004). , 795-802.

Essential Concepts of Oceanography

[141] Wada, E, & Hattori, A. *Nitrogen isotope effect in the assimilation of inorganic nitrogenous compounds by marine diatoms*. Geomicrobiology Journal, (1978)., 85-101.

[142] Montoya, J. P, & Mccarthy, J. J. *Isotopic fractionation during nitrate uptake by phytoplankton grown in continuous culture*. Journal of Plankton Research, (1995)., 439-464.

[143] Granger, J, et al. *Coupled nitrogen and oxygen isotope fractionation of nitrate during assimilation by cultures of marine phytoplankton*. Limnology and Oceanography, (2004)., 1763-1773.

[144] Hoch, M. P, Fogel, M. L, & Kirchman, D. L. *Isotope fractionation associated with ammonium uptake by a marine bacterium*. Limnology and Oceanography, (1992)., 1447-1459.

[145] Waser, N. A, et al. *Nitrogen isotope fractionation during nitrate, ammonium and urea uptake by marine diatoms and coccolithophores under various conditions of N availability*. Marine Ecology-Progress Series, (1998)., 29-41.

[146] Antia, N. J, Harrison, P. J, & Oliveira, L. *The role of dissolved organic nitrogen in phytoplankton nutrition, cell biology and ecology*. Phycologia, (1991)., 1-89.

[147] Eppley, R. W, & Thomas, W. H. *Comparison of half-saturation constants for growth and nitrate uptake of marine phytoplankron* Journal of Phycology, (1969)., 375-379.

[148] Strous, M, Kuenen, J. G, & Jetten, M. S. M. *Key physiology of anaerobic ammonium oxidation*. Applied and Environmental Microbiology, (1999)., 3248-3250.

[149] Lam, P, et al. *Revising the nitrogen cycle in the Peruvian oxygen minimum zone*. Proceedings of the National Academy of Sciences of the United States of America, (2009)., 4752-4757.

[150] Codispoti, L. A, et al. *The oceanic fixed nitrogen and nitrous oxide budgets: Moving targets as we enter the anthropocene?* Scientia Marina, (2001)., 85-105.

[151] Zumft, W. G, & Korner, H. *Enzyme diversity and mosaic gene organization in denitrification*. Antonie Van Leeuwenhoek International Journal of General and Molecular Microbiology, (1997)., 43-58.

[152] Codispoti, L. A, Yoshinari, T, & Devol, A. H. *Suboxic respiration in the oceanic water column*., in *Respiration in Aquatic Ecosystems*, P.A. del Giorgio and P.J.L.B. Williams, Editors. (2005). Blackwell Scientific: Oxford., 225-247.

[153] Zumft, W. G. *Cell biology and molecular basis of denitrification*. Microbiology and Molecular Biology Reviews, (1997)., 533-616.

[154] Ferguson, S. J. *Denitrification and its control*. Antonie Van Leeuwenhoek International Journal of General and Molecular Microbiology, (1994)., 89-110.

[155] Güven, D, et al. *Propionate oxidation by and methanol inhibition of anaerobic ammonium oxidizing bacteria* Applied and Environmental Microbiology, (2005)., 1066-1071.

[156] Kalvelage, T, et al. *Oxygen sensitivity of anammox and coupled N-cycle processes in oxygen minimum zones.* Plos One, (2011). , e29299.

[157] Jetten, M. S. M, et al. *Anaerobic ammonium oxidation by marine and freshwater planctomycete-like bacteria.* Applied Microbiology and Biotechnology, (2003). , 107-114.

[158] Van Niftrik, L, et al. *The anammoxasome: an intracytoplasmic compartment in anammox bacteria.* FEMS Microbiology Letters, (2004). , 7-13.

[159] Strous, M, et al. *Deciphering the evolution and metabolism of an anammox bacterium from a community genome.* Nature, (2006). , 790-794.

[160] Kartal, B, et al. *Anammox bacteria disguised as denitrifiers: nitrate reduction to dinitrogen gas via nitrite and ammonium.* Environmental Microbiology, (2007). , 635-642.

[161] Kuypers, M. M. M, et al. *Anaerobic ammonium oxidation by anammox bacteria in the Black Sea.* Nature, (2003). , 608-611.

[162] Dale, O. R, Tobias, C. R, & Song, B. *Biogeographical distribution of diverse anaerobic ammonium oxidizing (anammox) bacteria in Cape Fear River estuary.* Environmental Microbiology, (2009). , 1194-1207.

[163] Thamdrup, B, & Dalsgaard, T. *Production of N_2 through anaerobic ammonium oxidation coupled to nitrate reduction in marine sediments.* Applied and Environmental Microbiology, (2002). , 1312-1318.

[164] Dalsgaard, T, et al. *N_2 production by the anammox reaction in the anoxic water column of Golfo Dulce, Costa Rica.* Nature, (2003). , 606-608.

[165] Podlaska, A, et al. *Microbial community structure and productivity in the oxygen minimum zone of the eastern tropical North Pacific.* Deep Sea Research Part I: Oceanographic Research Papers, (2012). , 77-89.

[166] Hu, B, et al. *New anaerobic, ammonium-oxidizing community enriched from peat soil.* Applied and Environmental Microbiology, (2011). , 966-971.

[167] Humbert, S, et al. *Molecular detection of anammox bacteria in terrestrial ecosystems: distribution and diversity.* ISME J, (2009). , 450-454.

[168] Zhu, G, et al. *Anaerobic ammonia oxidation in a fertilized paddy soil.* ISME J, (2011). , 1905-1912.

[169] Erler, D. V, Eyre, B. D, & Davison, L. *The contribution of anammox and denitrification to sediment N_2 production in a surface flow constructed Wetland.* Environmental Science & Technology, (2008). , 9144-9150.

[170] Schubert, C. J, et al. *Anaerobic ammonium oxidation in a tropical freshwater system (Lake Tanganyika).* Environmental Microbiology, (2006). , 1857-1863.

[171] Hannig, M, et al. *Shift from denitrification to anammox after inflow events in the central Baltic Sea.* Limnology and Oceanography, (2007). , 1336-1345.

[172] Strous, M, et al. *The sequencing batch reactor as a powerful tool for the study of slowly growing anaerobic ammonium-oxidizing microorganisms.* Applied Microbiology and Biotechnology, (1998). , 589-596.

[173] van de GraafA.A., et al., *Autotrophic growth of anaerobic ammonium-oxidizing micro-organisms in a fluidized bed reactor.* Microbiology, (1996). , 2187-2196.

[174] Bock, E, et al. *Nitrogen loss caused by denitrifyling Nitrosomonas cells using ammonium or hydrogen as electron-donors and nitrite as electron-acceptor.* Archives of Microbiology, (1995). , 16-20.

[175] Blackmer, A. M, Bremner, J. M, & Schmidt, E. L. *Production of nitrous oxide by ammonia oxidizing chemoautotrophic microorganisms in soil.* Applied and Environtal Microbiology, (1980). , 1060-1066.

[176] Goreau, T. J, et al. *Production of 2-andN$_2$O by nitrifying bacteria at reduced concentrations of oxygen.* Applied and Environmental Microbiology, (1980). , 526-532.

[177] Bohlen, L, et al. *Benthic nitrogen cycling traversing the Peruvian oxygen minimum zone.* Geochimica et Cosmochimica Acta, (2011). , 6094-6111.

[178] Jensen, M. M, et al. *Intensive nitrogen loss over the Omani Shelf due to anammox coupled with dissimilatory nitrite reduction to ammonium.* Isme Journal, (2011). , 1660-1670.

[179] Rysgaard, S, & Risgaard, N. Petersen, and N.P. Sloth, *Nitrification, denitrification, and nitrate ammonification in sediments of two coastal lagoons in Southern France.* Hydrobiologia, (1996). p. , 133-141.

[180] Farías, L. *Potential role of bacterial mats in the nitrogen budget of marine sediments: the case of Thioploca spp.* Marine ecology progress series, (1998). , 291-292.

[181] Fossing, H, et al. *Concentration and transport of nitrate by the mat-forming sulphur bacterium Thioploca.* Nature, (1995). , 713-715.

[182] Schulz, H. N, et al. *Dense populations of a giant sulfur bacterium in Namibian shelf sediments.* Science, (1999). , 493-495.

[183] Cline, J. D, & Richards, F. A. *Oxygen deficient conditions and nitrate reduction in Eastern Tropical North-Pacific Ocean.* Limnology and Oceanography, (1972). , 885-900.

[184] Morrison, J. M, et al. *The oxygen minimum zone in the Arabian Sea during 1995.* Deep-Sea Research Part II-Topical Studies in Oceanography, (1999). , 1903-1931.

[185] Koeve, W, & Kähler, P. *Heterotrophic denitrification vs. autotrophic anammox- quantifying collateral effects on the oceanic carbon cycle.* Biogeosciences, (2010). , 2327-2337.

[186] Voss, M, & Montoya, J. P. *Nitrogen cycle. Oceans apart.* Nature, (2009). , 49-50.

[187] Bulow, S. E, et al. *Denitrification exceeds anammox as a nitrogen loss pathway in the Arabian Sea oxygen minimum zone.* Deep-Sea Research Part I-Oceanographic Research Papers, (2010). , 384-393.

[188] Galan, A, et al. *Anammox bacteria and the anaerobic oxidation of ammonium in the oxygen minimum zone off northern Chile.* Deep-Sea Research Part II-Topical Studies in Oceanography, (2009). , 1125-1135.

[189] Devol, A. H, et al. *Denitrification rates and excess nitrogen gas concentrations in the Arabian Sea oxygen deficient zone.* Deep-Sea Research Part I-Oceanographic Research Papers, (2006). , 1533-1547.

[190] Van Mooy, B. A. S, Keil, R. G, & Devol, A. H. *Impact of suboxia on sinking particulate organic carbon: Enhanced carbon flux and preferential degradation of amino acids via denitrification.* Geochimica et Cosmochimica Acta, (2002). , 457-465.

[191] Brandes, J. A, et al. *Isotopic composition of nitrate in the central Arabian Sea and eastern tropical North Pacific: A tracer for mixing and nitrogen cycles.* Limnology and Oceanography, (1998). , 1680-1689.

[192] Sigman, D. M, et al. *The $\delta^{15}N$ of nitrate in the Southern Ocean: Consumption of nitrate in surface waters.* Global Biogeochemical Cycles, (1999). , 1149-1166.

[193] Voss, M, Dippner, J. W, & Montoya, J. P. *Nitrogen isotope patterns in the oxygen-deficient waters of the Eastern Tropical North Pacific Ocean.* Deep-Sea Research Part I-Oceanographic Research Papers, (2001). , 1905-1921.

[194] Cline, J. D, & Kaplan, I. R. *Isotopic fractionation of dissolved nitrate during denitrification in the eastern tropical north Pacific Ocean.* Marine Chemistry, (1975). , 271-299.

[195] Lehmann, M. F, et al. *The distribution of nitrate $^{15}N/^{14}N$ in marine sediments and the impact of benthic nitrogen loss on the isotopic composition of oceanic nitrate.* Geochimica et Cosmochimica Acta, (2007). , 5384-5404.

[196] Codispoti, L. A. *An oceanic fixed nitrogen sink exceeding 400 Tg Na-1 vs the concept of homeostasis in the fixed nitrogen inventory.* Biogeosciences, (2007). , 233-253.

[197] Codispoti, L. A, & Christensen, J. P. *Nitrification, denitrification and nitrous-oxide cycling in the eastern tropical South-Pacific Ocean.* Marine Chemistry, (1985). , 277-300.

[198] Brandes, J. A, & Devol, A. H. *A global marine-fixed nitrogen isotopic budget: Implications for Holocene nitrogen cycling.* Global Biogeochemical Cycles, (2002). , 1120-1134.

[199] Galloway, J. N, et al. *Nitrogen Cycles: Past, Present, and Future.* Biogeochemistry, (2004). , 153-226.

[200] Gruber, N. *The marine nitrogen cycle: Overview and challanges,* in *Nitrogen in Marine Environment,* D.G. Capone, et al., Editors. (2008). Elsevier Inc. , 1-51.

[201] Codispoti, L. A. *Interesting times for marine N_2O.* Science, (2010). , 1339-1340.

[202] Galbraith, E. D, et al. *Global nitrogen isotopic constraints on the acceleration of oceanic denitrification during deglacial warming.* Nature Geosciences, (2012). submitted.

[203] Altabet, M. A. *Constraints on oceanic N balance/imbalance from sedimentary ^{15}N records.* Biogeosciences, (2007). , 75-86.

[204] Somes, C. *Nitrogen Isotopes in the Global Ocean: Constraints on the pre-industrial fixed-N budget,* in *Biogeochemical modelling* (2012). Christian-Albrechts-Universität zu Kiel: Kiel.

Chapter 2

A Statistical Approach for Wave-Height Forecast Based on Spatiotemporal Variation of Surface Wind

Tsukasa Hokimoto

Additional information is available at the end of the chapter

1. Introduction

For accurate wave-height forecasts, it is necessary to take into account changes in various physical phenomena related to meteorology, because wave motion is affected by changes in ocean wind. However, it is generally difficult to carry out continuous field measurements of such physical phenomena in an area of investigation at sea, because of the lack of facilities required for such measurements. The physical processes related to meteorological or oceanographic phenomena are thought to have changeable correlations in space and time. Therefore, perhaps it is possible to forecast wave-height changes effectively by developing a method that takes spatiotemporal features into consideration. The Japan Meteorological Agency has set up regional stations for ground-based meteorological monitoring of coastal areas using ultrasonic wave-height meters. The systems is referred to as the Automated Meteorological Data Acquisition System (AMeDAS). An approach for wave-height forecast, based on spatiotemporal wind motions monitored at multiple ground-based AMeDAS stations, provides an alternative method for solving the above measurement problem.

One of traditional approaches for analyzing wave-height changes is to regard sea surface oscillations to be a probabilistic phenomenon and then to consider statistical approaches for expressing the dynamics of wave heights from this standpoint. Statistical models for dealing with measurements of long-term variations in wave height have been considered mainly from two perspectives: nonstationarity (e.g., Scheffner and Borgman (1992), Athanassoulis and Stefanakos (1995), Guedes Soares and Ferreira (1996)) and nonlinearity (e.g., Scotto and Guedes Soares (2000)). On the other hand, statistical methods for modeling wave height that take into account changes in wind speed and wind direction have also been considered (e.g., Hokimoto and Shimizu (2008), Hokimoto (2012)). However, adequate statistical considerations of whether or not the use of spatiotemporal wind motion is an effective method for expressing and forecasting changes in wave height have not yet been undertaken.

Also, it is not clear that statistical spatiotemporal models can improve forecasting accuracy when traditional statistical models are used.

In this chapter, we consider the points above through the development of a statistical spatiotemporal model. We first consider a time series model for expressing the relationship between wave-height changes measured in a coastal area and wind motion (i.e., wind direction and wind speed) measured at a single meteorological AMeDAS station, by extending the model considered in Hokimoto and Shimizu (2008). Then we propose a method to take spatiotemporally measured wind motion data into account, by extending the model structure developed above. Also, the applicability of the method for the analysis of actual phenomena is evaluated by a case study of wave-height forecast from a coastal area of Hokkaido, Japan.

This chapter is organized as follows. In Section 2, we describe field measurements of wave height and wind motion, including a preliminary statistical analysis of the measured data. In Section 3, we develop a statistical spatiotemporal model for forecasting wave height. The effectiveness of the method is examined by forecasting experiments in Section 4. Then in Section 5, we show the applicability of the method in the analysis of actual phenomena through a case study. Conclusions are given in Section 6.

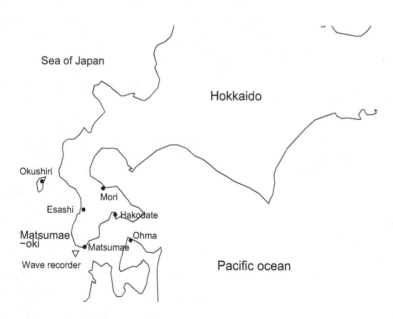

Figure 1. Locations of wave recorder and meteorological stations

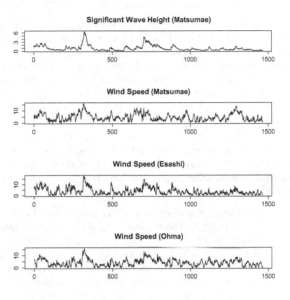

Figure 2. Wave height at Matsumae-oki and wind speeds over Matsumae, Esashi and Ohma

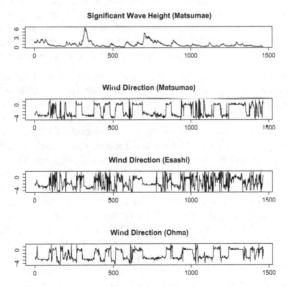

Figure 3. Wave height at Matsumae-oki and wind directions over Matsumae, Esashi and Ohma

2. Monitoring wave-height and spatiotemporal changes of wind motion

In this chapter, we consider a case study of wave-height forecast as monitored at Matsumae-oki, Hokkaido, Japan. Matsumae is a famous fisheries town where many of the local people are involved with various activities related to the sea as part of their daily lives. The neighboring area of Matsumae-oki, directly on the coast, is known for its dangerous seas. Sea conditions here quickly tend to become rough after the onset of ocean winds. In this region, wave observations using an ultrasonic wave-height metering system have been undertaken by the Japan Meteorological Agency since 1979. The sensor of the wave-height meter is located on the seabed at a depth of approximately 50 m at $42°24'38''$ N, $140°05'50''$ E. The meter measures the movement of the sea surface above using ultrasonic waves, then transmits a stream of data, including significant wave heights and wave period, to a main recording center. At the AMeDAS stations located in the neighboring area of Matsumae-oki, atmospheric properties such as temperature, precipitation, wind speed, wind direction and hours of sunlight are also measured and sent to the main center.

Figure 1 displays the location of the wave recorder and six towns in the neighboring area where AMeDAS stations are located. Also Figures 2 and 3 show an example of the measured data on 1/3 significant wave height (m), wind speed (m/s) and wind direction (rad.), for the period from April to May of 2010, where the sampling time interval is 1 hour. As noted in Figure 3, the origin of the wind direction is to the east and a positive increase corresponds to a clockwise change in direction. Since the AMeDAS station at Matsumae is closest to the wave recorder, it may be possible to use the wind motion monitored at Matsumae only, to develop a wave-height forecasting method. However, it is unclear whether this method is reasonable for expressing the dynamic structure of wave-height changes. In Figure 2, for example, it appears that the characteristics of wave-height change at Matsumae-oki are less synchronous with the wind speed changes at Matsumae, which is closest to the wave recorder, than that at Esashi, which is farther away than Matsumae. From a physical standpoint, this phenomenon can be explained by the interruption of wind flow by geographical features such as mountains.

We are interested in whether or not taking into account the spatiotemporal structure of the measured data on surface wind, monitored from the multiple AMeDAS stations, affects the forecast of wave heights. To investigate the effectiveness of introducing the class of spatiotemporal models, we have undertaken a preliminary analysis of the spatial and temporal correlation structure. Let $\{WH_t\}$, $\{WS_t\}$ and $\{WD_t\}$ be time series on 1/3 significant wave height, wind speed and wind direction, respectively. Figure 4 displays the cross correlation between the differenced time series $\{\nabla(WS_t \cos(WD_t))\}$ and $\{\nabla WH_t\}$ based on the measured data at Matsumae, Esashi and Ohma, for the three periods [150-250] (top), [300-400] (middle) and [450-550] (bottom). Here, the two dotted lines correspond to Bartlett's bounds to test the significance of the correlation between the two time series. Matsumae, Esashi and Ohma are located to the northeast, north and east of the measuring point, respectively. In [150-250], the wave-height change is most closely correlated with the wind motion over Matsumae, but the correlation gradually reduces over time. In fact, the town that gives the maximum cross-correlation value changes from Matsumae to Esashi, then to Ohma. This suggests the possibility of taking the contribution of spatiotemporal structure into account to improve the accuracy of wave-height forecasts.

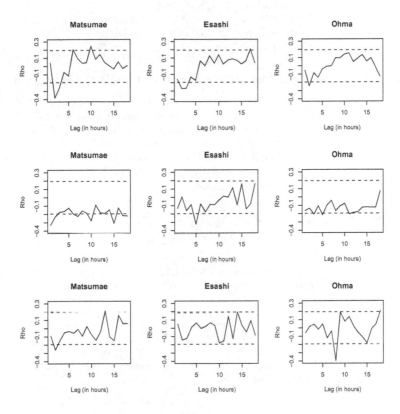

Figure 4. Change in cross correlation coefficients between $\{\nabla(WS_t \cos(WD_t))\}$ and $\{\nabla WH_t\}$ (from the top, estimated results for the periods [150-250], [300-400] and [450-500])

3. A statistical method for forecasting wave-height changes from spatiotemporal wind motion

In this section, we follow on from the result of the preliminary analysis presented in the previous section by considering the statistical spatiotemporal modeling of wave weight. As shown in Figures 2 and 3, the characteristics of the wind speed and wind direction time series are different. For this reason, different classes of time series models have been considered to express changes in wind speed and wind direction. For wind speed, linear models such as ARMA (e.g., Philippopoulos and Deligiorgi (2009)) and GARCH (e.g., Tol (1997) and Liu et al. (2011)) have been applied for analysis. In contrast, wind direction time series frequently tend to show rapid changes, which have different characteristics than those of wind speed. Johnson and Wehrly (1978) considered a linear regression model to deal with directional data and Hokimoto and Shimizu (2008) considered a time series model for the situation above. We extend the model structure of Hokimoto and Shimizu (2008) to consider the spatiotemporal relationship between wind motion and wave height. We first consider a time series model to forecast wave height based on the wind motion monitored at a single meteorological station.

Then, we extend the model so that it is applicable to the spatiotemporal wind speed and wind direction data. Our goal is to develop a predictor of wave height, WH_{T+l} ($l = 1, \ldots, L$), based on the measured data $\{WH_t\}$, $\{WS_t\}$ and $\{WD_t\}$ ($t = 1, \ldots, T$).

3.1. Forecasting wave height from surface wind over a single meteorological station

A good place to start a consideration of wave-height modeling is from forecasts of wave height from a single meteorological station. We assume that $\{WD_t\}$ ($-\pi \leq WD_t \leq \pi$) follows a von Mises process of the first order, as considered by Breckling (1989). Under this assumption, the conditional distribution under $\{WD_{t-1}\}$ is observed to follow a von Mises distribution with a mean $\mu_{(WD),t}$ and concentration $\rho_{(WD),t}$, which satisfy

$$\rho_{(WD),t} \begin{pmatrix} \cos(\mu_{(WD),t}) \\ \sin(\mu_{(WD),t}) \end{pmatrix} = k_1 \begin{pmatrix} \cos(WD_{t-1}) \\ \sin(WD_{t-1}) \end{pmatrix} + k_0 \begin{pmatrix} 1 \\ 0 \end{pmatrix}$$

$$k_0 > 0, \ k_1 > 0, \ -\pi \leq \mu_{(WD),t} \leq \pi, \ \rho_{(WD),t} > 0$$

where k_0 and k_1 are unknown parameters which take positive values. The conditional probability density function of WD_t, under WD_{t-1} is observed, can be written by

$$f(WD_t|WD_{t-1}) = \frac{1}{2\pi I_0(\rho_{(WD),t})} \exp\{k_1 \cos(WD_t - WD_{t-1}) + k_0 \cos WD_t\} \tag{1}$$

where $I_0(\cdot)$ is a modified zero-order Bessel function. (1) can be rewritten by the probability density function of the von Mises distribution

$$f(WD_t|WD_{t-1}) = \frac{1}{2\pi I_0(\rho_{(WD),t})} \exp(\rho_{(WD),t} \cos(WD_t - \mu_{(WD),t}))$$

where

$$\mu_{(WD),t} = \tan^{-1}\left(\frac{k_1 \sin(WD_{t-1})}{k_1 \cos(WD_{t-1}) + k_0}\right) \tag{2}$$

and

$$\rho_{(WD),t} = \sqrt{(k_1 \cos(WD_{t-1}) + k_0)^2 + (k_1 \sin(WD_{t-1}))^2} \tag{3}$$

which means that the parameters ($\mu_{(WD),t}$, $\rho_{(WD),t}$) change depending on (k_0, k_1) and WD_{t-1}. When both k_0 and k_1 are sufficiently small, the concentration parameter $\rho_{(WD),t}$ also becomes small and therefore (1) can be approximated by a uniform distribution. Conversely, when k_0 or k_1 becomes larger, $\rho_{(WD),t}$ also gets larger and changes to a distribution which is concentrated around $\mu_{(WD),t}$.

Next, consider a method for the estimation of the above process. To guarantee the positivity of $\rho_{(WD),t}$, we write $k_i = \exp(c_i)$ $(i = 0, 1)$ and then estimate values of the parameter (c_0, c_1). Suppose that the conditional density function of WD_t can be written by

$$f(WD_t|WD_{t-1}, \ldots, WD_1) = f(WD_t|WD_{t-1}).$$

Then the likelihood function $f(WD_1, \ldots, WD_T)$ can be written as

$$\prod_{j=2}^{T} \frac{1}{2\pi I_0(\rho_{(WD),j})} \exp\{\exp(c_1)\cos(WD_j - WD_{j-1}) + \exp(c_0)\cos(WD_j)\} \cdot f(WD_1). \quad (4)$$

(4) is a function of the parameters (c_0, c_1) only, and their values can be estimated by maximization of (4). Let $(\hat{c}_0, \hat{c}_1)$ be the maximum likelihood estimates obtained above. Then $\hat{\mu}_{(WD),t}$ and $\hat{\rho}_{(WD),t}$ can be estimated respectively by

$$\hat{\mu}_{(WD),t} = \tan^{-1}\left(\frac{\exp(\hat{c}_1)\sin(WD_{t-1})}{\exp(\hat{c}_1)\cos(WD_{t-1}) + \exp(\hat{c}_0)}\right) \quad (5)$$

and

$$\hat{\rho}_{(WD),t} = \sqrt{(\exp(\hat{c}_1)\cos(WD_{t-1}) + \exp(\hat{c}_0))^2 + (\exp(\hat{c}_1)\sin(WD_{t-1}))^2}. \quad (6)$$

Figure 5 shows a time series of the estimated value of $\{\cos(\hat{\mu}_{(WD),t})\}$, its autocorrelation and the time series $\{\hat{\rho}_{(WD),t}\}$. Note that $\{\cos(\hat{\mu}_{(WD),t})\}$ has a tendency to change in a certain range with a significant autocorrelation, and $\{\hat{\rho}_{(WD),t}\}$ exhibits nonstationarity in the sense that the mean and variance change clearly over time. Therefore, $\{\cos(\hat{\mu}_{(WD),t})\}$ and $\{\hat{\rho}_{(WD),t}\}$ can be regarded as stationary and nonstationary time series, respectively.

The modeling strategy applied here is to extend the nonstationary model structure considered by Hokimoto and Shimizu (2008) so that the new model can take into account the synchronous relationship between the von Mises process assumed for the change in wind direction. The extended model is written by

$$\nabla WH_t = \sum_{i=1}^{p} \alpha_i^{(1)} \nabla WH_{t-i} + \sum_{i=1}^{p}\sum_{k=1}^{K} \beta_{i,k}^{(1)} \nabla(\rho_{(WD),t-i} WS_{t-i}\cos(kWD_{t-i}))$$

$$+ \sum_{i=1}^{p}\sum_{k=1}^{K} \gamma_{i,k}^{(1)} \nabla(\rho_{(WD),t-i} WS_{t-i}\sin(kWD_{t-i})) + \sum_{i=1}^{p} \delta_i^{(1)}\cos(\mu_{(WD),t-i})$$

$$+ \sum_{i=1}^{p} \omega_i^{(1)}\sin(\mu_{(WD),t-i}) + \varepsilon_t^{(1)}, \quad \varepsilon_t^{(1)} \sim WN(0, \sigma_{1,h}^2)$$

where p and K are orders, $(\alpha^{(1)}, \beta^{(1)}, \gamma^{(1)}, \delta^{(1)}, \omega^{(1)})$ are unknown constants and $\varepsilon_t^{(1)}$ is a random variable that follows a zero-mean white noise process.

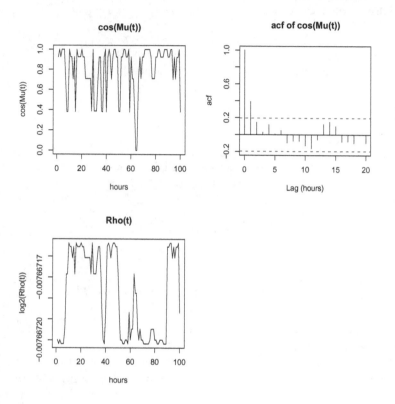

Figure 5. Time series $\{\cos(\hat{\mu}_{(WD),t})\}$ and its autocorrelation (top) and time series $\{\hat{\rho}_{(WD),t}\}$ (bottom)

Similarly, we write $\nabla(\rho_{(WD),t} WS_t \cos(hWD_t))$ and $\nabla(\rho_{(WD),t} WS_t \sin(hWD_t))$ $(h = 1,\ldots,K)$ in the form

$$\nabla(\rho_{(WD),t} WS_t \cos(hWD_t)) = \sum_{i=1}^{p} \alpha_i^{(2)} \nabla WH_{t-i} + \sum_{i=1}^{p}\sum_{k=1}^{K} \beta_{i,k}^{(2)} \nabla(\rho_{(WD),t-i} WS_{t-i} \cos(kWD_{t-i}))$$

$$+ \sum_{i=1}^{p}\sum_{k=1}^{K} \gamma_{i,k}^{(2)} \nabla(\rho_{(WD),t-i} WS_{t-i} \sin(kWD_{t-i})) + \sum_{i=1}^{p} \delta_i^{(2)} \cos(\mu_{(WD),t-i})$$

$$+ \sum_{i=1}^{p} \omega_i^{(2)} \sin(\mu_{(WD),t-i}) + \varepsilon_t^{(2)}, \quad \varepsilon_t^{(2)} \sim WN(0,\sigma_{2,h}^2)$$

and $\cos(\mu_{(WD),t})$ and $\sin(\mu_{(WD),t})$ as

$$\cos(\mu_{(WD),t}) = \sum_{i=1}^{p} \alpha_i^{(3)} \nabla WH_{t-i} + \sum_{i=1}^{p} \sum_{k=1}^{K} \beta_{i,k}^{(3)} \nabla(\rho_{(WD),t-i} WS_{t-i} \cos(kWD_{t-i}))$$

$$+ \sum_{i=1}^{p} \sum_{k=1}^{K} \gamma_{i,k}^{(3)} \nabla(\rho_{(WD),t} WS_{t-i} \sin(kWD_{t-i})) + \sum_{i=1}^{p} \delta_i^{(3)} \cos(\mu_{(WD),t-i})$$

$$+ \sum_{i=1}^{p} \omega_i^{(3)} \sin(\mu_{(WD),t-i}) + \varepsilon_{t-i}^{(3)}, \quad \varepsilon_t^{(3)} \sim WN(0, \sigma_{3,h}^2).$$

A state vector at time t is defined by

$$y_t^{(K)} \equiv (\nabla WH_t, \ \nabla(\rho_{(WD),t} WS_t \cos(WD_t)), \ \nabla(\rho_{(WD),t} WS_t \sin(WD_t)), \ \dots,$$

$$\nabla(\rho_{(WD),t} WS_t \cos(K \cdot WD_t)), \ \nabla(\rho_{(WD),t} WS_t \sin(K \cdot WD_t)), \ \cos(\mu_{(WD),t}), \ \sin(\mu_{(WD),t}))' \quad (7)$$

Then, the models given above can be unified as a multivariate AR model

$$y_t^{(K)} = A_1^{(K)} y_{t-1}^{(K)} + \cdots + A_p^{(K)} y_{t-p}^{(K)} + \delta_t^{(K)}, \qquad \delta_t^{(K)} \sim WN(0, \Sigma^{(K)}) \quad (8)$$

where $A_i^{(K)}$ $(i = 1, \dots, p)$ is an unknown coefficients matrix.

The predictor of (8) can be constructed in the following way. We first estimate (c_0, c_1) by maximizing the likelihood of (4) and then obtain the values of $\{\hat{\mu}_{(WD),t}\}$ and $\{\hat{\rho}_{(WD),t}\}$ by (5) and (6), respectively. Next, we construct the sequence of $y_t^{(K)}$ by

$$y_t^{(K)} \equiv (\nabla WH_t, \ \nabla(\hat{\rho}_{(WD),t} WS_t \cos(WD_t)), \ \nabla(\hat{\rho}_{(WD),t} WS_t \sin(WD_t)), \ \dots,$$

$$\nabla(\hat{\rho}_{(WD),t} WS_t \cos(K \cdot WD_t)), \ \nabla(\hat{\rho}_{(WD),t} WS_t \sin(K \cdot WD_t)), \ \cos(\hat{\mu}_{(WD),t}), \ \sin(\hat{\mu}_{(WD),t}))'$$

and then fit (8) to $\{y_t^{(K)}\}$ $(t = 1, \dots, T)$. A linear predictor based on (8) can be constructed by

$$\hat{y}_{T+l}^{(K)} = \hat{A}_1^{(K)} z_{T+l-1}^{(K)} + \hat{A}_2^{(K)} z_{T+l-2}^{(K)} + \cdots + \hat{A}_p^{(K)} z_{T+l-p}^{(K)}, \quad (9)$$

$$z_{T+l-m}^{(K)} = y_{T+l-p}^{(K)} (l \le p), \qquad z_{T+l-m}^{(K)} = \hat{y}_{T+l-p}^{(K)} (l > p)$$

where $\hat{A}_i^{(K)}$ $(i = 1, \dots, p)$ are the least squares estimator (e.g., Brockwell and Davis (1996)).

3.2. Forecasting wave height from spatiotemporal surface wind over multiple meteorological stations

In this subsection, we extend the method presented in 3.1 so that it is applicable to the spatiotemporal data measured at multiple meteorological stations. We continue our consideration by setting $K = 1$. Here we consider a spatiotemporal model by expressing the situation that the wind flow, which has the largest impact on wave height, changes over time. First, rather than (7), the state vector for fitting the multivariate AR model (8) is defined by

$$\boldsymbol{y}_{t|s^*} \equiv (\nabla WH_t, \ \nabla(\rho_t^{(s^*)} WS_t^{(s^*)} \cos(WD_t^{(s^*)})), \ \nabla(\rho_t^{(s^*)} WS_t^{(s^*)} \sin(WD_t^{(s^*)})), \ \cos(\mu_{(WD),t}^{(s^*)}), \ \sin(\mu_{(WD),t}^{(s^*)}))'$$

where s^* means the meteorological station that measures the wind with the largest impact on wave-height change at s ($s = 1, \ldots, 6$) meteorological stations, and $WS_t^{(s^*)}$ means WS_t at the station.

The value of s^* is chosen by a statistical method based on measured data. It is defined as the value of s which minimizes the mean squared errors on forecasts made one-step ahead. Let $WS_t^{(s)}$ and $WD_t^{(s)}$ be wind speed and wind direction data, respectively, measured at the sth meteorological station. We first obtain forecasts of WH_t, one-step ahead in time, based on $\{WS_t^{(s)}\}$ and $\{WD_t^{(s)}\}$ ($s = 1, \ldots, 6$), say $\widetilde{WH}_{t+1}^{(s)}$ ($t = 1, \ldots, T - 1$). The forecasts are obtained by fitting (8) to the sequence constructed by

$$\boldsymbol{y}_{t|s} = (\nabla WH_t, \ \nabla(\rho_t^{(s)} WS_t^{(s)} \cos(WD_t^{(s)})), \ \nabla(\rho_t^{(s)} WS_t^{(s)} \sin(WD_t^{(s)})), \ \cos(\mu_{(WD),t}^{(s)}), \ \sin(\mu_{(WD),t}^{(s)}))'.$$

Then, we choose the value of s^* by

$$s^* = \arg\min_{1 \leq s \leq 6} R(s, \tau(s))$$

where

$$R(s, \tau(s)) = \frac{1}{\tau(s)} \sum_{t=T-\tau(s)+1}^{T} (WH_t - \widetilde{WH}_t^{(s)})^2$$

where $\tau(s)$ means the local time interval, which depends on s. Thus, the forecasted value of WH_{T+l} ($l = 1, \ldots, L$) can be obtained by applying the predictor (9) to the sequence $\{\boldsymbol{y}_{t|s^*}; t = 1, \ldots, T\}$.

4. Evaluation of forecasting accuracies based on numerical experiments

An examination of the applicability of the spatiotemporal model, presented in the previous section, is required from the standpoint of forecasting accuracy. For this purpose, we carried out a forecasting experiment for the significant wave height by using several statistical models, and then compared forecasting accuracies among the models. The experimental procedure was as follows. First, the model was fit to the time series data of 100 samples (i.e., 100 hours) and forecasted values up to five steps ahead (i.e., 5 hours ahead) were determined. Next, the time point used for the starting forecast was changed randomly and the forecasting step above was repeated. After repeating the procedure, we obtained mean squared errors (MSE) and calculated the correlation between forecasted and actual values (COR), based on forecasted values and measured data. At the same time, the MSEs and CORs were also obtained using several traditional nonstationary time series models in a similar way. These values were compared between models to determine the class of models that gives the best forecasting accuracy. When we fit the models above, the order was determined by the Akaike Information Criterion (AIC).

4.1. Forecasting accuracy based on the surface wind monitored at a single meteorological station

We first investigate the case of forecasts based on the wind motions measured at Matsumae, the nearest location to the wave recorder. For this purpose, we obtained forecasted values of wave heights using wind speed and wind direction data measured at Matsumae. The statistical models introduced for comparisons are as follows.

(i) $WH_t = \sum_{i=1}^{p} \alpha_i WH_{t-i} + \varepsilon_{1,t}, \quad \varepsilon_{1,t} \sim WN(0, \sigma_1^2)$

(ii) $\nabla WH_t = \sum_{i=1}^{p} \beta_i \nabla WH_{t-i} + \varepsilon_{2,t}, \quad \varepsilon_{2,t} \sim WN(0, \sigma_2^2)$

(iii) $y_t = A_1 y_{t-1} + \cdots + A_p y_{t-p} + \delta_t, \quad \delta_t \sim WN(0, \Sigma),$
$y_t = (\nabla WH_t, \nabla WS_t)'$

(iv) $y_t = A_1 y_{t-1} + \cdots + A_p y_{t-p} + \delta_t, \quad \delta_t \sim WN(0, \Sigma),$
$y_t = (\nabla WH_t, \nabla(WS_t \cos(WD_t)))'$

(v) $y_t = A_1 y_{t-1} + \cdots + A_p y_{t-p} + \delta_t, \quad \delta_t \sim WN(0, \Sigma),$
$y_t = (\nabla WH_t, \nabla(\rho_t WS_t \sin(WD_t)), \sin(\mu_{(WD),t}))'$

(vi) $y_t = A_1 y_{t-1} + \cdots + A_p y_{t-p} + \delta_t, \quad \delta_t \sim WN(0, \Sigma),$
$y_t = (\nabla WH_t, \nabla(\rho_t WS_t \cos(WD_t)), \cos(\mu_{(WD),t}))'$

(i) and (ii) are models based on wave height only; the former is a stationary univariate AR(p) model and the latter is a nonstationary univariate ARIMA(p,1,0) model. (iii) is a nonstationary vector autoregressive model that takes into account changes in $\{WS_t\}$ and $\{WH_t\}$. Models (iv)-(vi) are nonstationary models that take into account both $\{WS_t\}$ and $\{WD_t\}$ as covariates. Note that (v) and (vi) belong to a class of model presented in the previous section.

Model	MSE					COR				
	1-step	2-step	3-step	4-step	5-step	1-step	2-step	3-step	4-step	5-step
(i)	0.039	0.089	0.141	0.195	0.224	0.982	0.955	0.927	0.886	0.862
(ii)	0.028	0.070	0.099	0.154	0.175	0.985	0.959	0.943	0.906	0.893
(iii)	0.030	0.072	0.101	0.153	0.166	0.984	0.960	0.944	0.912	0.903
(iv)	0.027	0.065	0.094	0.142	0.158	0.986	0.964	0.947	0.919	0.906
(v)	0.028	0.068	0.092	0.139	0.151	0.985	0.963	0.949	0.921	0.913
(vi)	0.027	0.064	0.092	0.140	0.157	0.986	0.965	0.950	0.919	0.908

Table 1. Forecasting accuracies of time series forecasts using each model

Table 1 shows the MSEs and CORs used with models (i)-(vi) for forecast up to five steps ahead, obtained by 130 repetitions. Comparison of (i) and (ii) shows that the forecast based on the nonstationary ARIMA(p,1,0) model is more accurate than a stationary AR model, which suggests that nonstationary models tends to give better forecasts than stationary models. Also, based on comparisons between (ii) and (iii) and between (ii) and (iv), we observe the tendency for (iii) and (iv) to give better forecasting accuracies than (ii), which also highlights the possibility of taking wind motion into account as a covariate contribution to improve forecasting accuracy. Comparisons between (iv) and (v) and between (iv) and (vi) suggest that a model which takes into account von Mises process on $\{WD_t\}$ will improve forecasting accuracies of (iii) and (iv), further. Based on the result above, it is determined that changes in the parameters of von Mises process assuming $\{WD_t\}$ tend to synchronize with $\{WS_t\}$ and $\{WH_t\}$.

4.2. Effect of spatiotemporal models on improvement of the forecasting accuracy

We next consider whether or not taking into account the wind motions measured at multiple meteorological stations contributes to the improvement of forecasting accuracies obtained in 4.1. To examine this point, we carried out forecasting experiments similar to those presented in 4.1. Additionally, we introduce the following spatiotemporal models for comparisons of forecasting accuracies.

vii) $\quad y_t = A_1 y_{t-1} + \cdots + A_p y_{t-p} + \delta_t, \quad \delta_t \sim WN(0, \Sigma),$
$\quad\quad y_t = (\nabla WH_t, \nabla(WS_t^{(1)} \cos(WD_t^{(1)})), \ldots, \nabla(WS_t^{(6)} \cos(WD_t^{(6)})))'$

viii) $\quad y_t = A_1 y_{t-1} + \cdots + A_p y_{t-p} + \delta_t, \quad \delta_t \sim WN(0, \Sigma),$
$\quad\quad y_t = (\nabla WH_t, \nabla(WS_t^{(s^*)} \cos(WD_t^{(s^*)})))'$

ix) $\quad y_t = A_1 y_{t-1} + \cdots + A_p y_{t-p} + \delta_t, \quad \delta_t \sim WN(0, \Sigma),$
$\quad\quad y_t = (\nabla WH_t, \nabla(\rho_t^{(s^*)} WS_t^{(s^*)} \sin(WD_t^{(s^*)})), \sin(\mu_{(WD),t}^{(s^*)}))'$

x) $\quad y_t = A_1 y_{t-1} + \cdots + A_p y_{t-p} + \delta_t, \quad \delta_t \sim WN(0, \Sigma),$
$\quad\quad y_t = (\nabla WH_t, \nabla(\rho_t^{(s^*)} WS_t^{(s^*)} \cos(WD_t^{(s^*)})), \cos(\mu_{(WD),t}^{(s^*)}))'$

Table 2 shows the MSEs and CORs obtained in the forecasting experiments above. (vii) is a standard vector autoregressive model based on multivariate wind speed and wind direction

Model	MSE					COR				
	1-step	2-step	3-step	4-step	5-step	1-step	2-step	3-step	4-step	5-step
(vii)	0.033	0.067	0.097	0.142	0.161	0.982	0.963	0.947	0.918	0.905
(viii)	0.027	0.067	0.094	0.136	0.153	0.986	0.965	0.952	0.927	0.917
(ix)	0.025	0.065	0.086	0.134	0.149	0.986	0.964	0.953	0.923	0.913
(x)	0.024	0.065	0.091	0.125	0.149	0.987	0.964	0.950	0.929	0.913

Table 2. Forecasting accuracies of spatiotemporal forecasts using each model

time series data measured at six meteorological stations. The results show that the forecasting accuracy of (vii) tends to become worse than forecasts based on a single meteorological station, as investigated in 4.1. This is likely because this class of model tends to have a large number of parameters that need to be estimated, which leads to negative impacts on forecasting accuracy. On the other hand, (viii) has fewer parameters than (vii), which leads to improved forecasting results as shown in Table 1. Furthermore, the class of the proposed models, (ix) and (x), contributes to the improvement of forecasting accuracies by (viii), which gives the best forecasting accuracy in our experiments.

5. Applying spatiotemporal modeling for wave-height forecasts

5.1. Robustness on wave-height forecasts over four seasons

In Japan, there are unique pressure pattern characteristics for each season, and it is therefore necessary to examine the applicability of the proposed model throughout the year. The purpose of this section is to investigate whether the model provides a robust forecast of wave height over the four seasons.

Figures 6-8 display, respectively, time series for the one-third significant wave height, wind speed and wind direction, in the spring (Apr. 1 - May 31), summer (Jul. 1-Aug. 31), autumn (Oct. 1 - Nov. 31) and winter (Jan. 1 - Feb.28), measured in Matsumae-oki and Matsumae. Overall, the characteristics of the changes are different for each season. Particularly in winter, the latent stochastic abilities differ from those in the other seasons, under a background of strong stable seasonal winds blowing from the northwest.

We carried out forecasting experiments similar to those presented in 4.1 and 4.2. In this experiment, some of the seven models introduced in 4.1 and 4.2 were adopted, and then their MSEs and CORs were compared for each season. More specifically, we adopted models (i)-(iv) from the time series models introduced in 4.1, and (vii), (viii) and (x) from the spatiotemporal models in 4.2.

Table 3 shows the MSEs obtained from forecasting experiments for each season. In spring, summer and autumn, the proposed spatiotemporal model (x) was evaluated as having an effective model structure for robust forecast, in the sense that it tends to give the best MSEs of the seven models tested. Note that, as for the winter forecast, there is no clear improvement on the MSEs. This tendency is presumed to result from the wind direction in this season, which is generally from the northwest with the degree of fluctuation that is smaller than in the other three seasons.

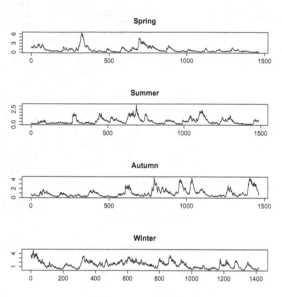

Figure 6. Changes in 1/3 significant wave height (m) for the four seasons (Matsumae-oki)

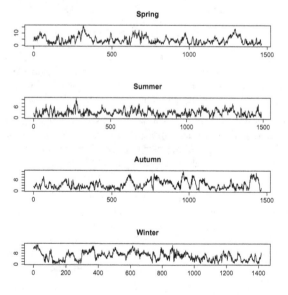

Figure 7. Changes in wind speed (m/s) for the four seasons (Matsumae)

A) Spring

Model	Region(s)	1-step	2-step	3-step	4-step	5-step
(i)	1	0.039	0.089	0.141	0.195	0.224
(ii)	1	0.028	0.070	0.099	0.154	0.175
(iii)	1	0.030	0.072	0.101	0.153	0.166
(iv)	1	0.027	0.065	0.094	0.142	0.158
(vii)	6	0.033	0.067	0.097	0.142	0.161
(viii)	6	0.027	0.067	0.094	0.136	0.153
(x)	6	0.024	0.065	0.091	0.125	0.150

B) Summer

Model	Region(s)	1-step	2-step	3-step	4-step	5-step
(i)	1	0.015	0.025	0.033	0.048	0.073
(ii)	1	0.011	0.015	0.023	0.040	0.059
(iii)	1	0.011	0.014	0.023	0.039	0.058
(iv)	1	0.011	0.014	0.022	0.038	0.056
(vii)	6	0.012	0.017	0.025	0.040	0.056
(viii)	6	0.012	0.014	0.022	0.036	0.054
(x)	6	0.012	0.016	0.022	0.036	0.053

C) Autumn

Model	Region(s)	1-step	2-step	3-step	4-step	5-step
(i)	1	0.024	0.091	0.139	0.222	0.267
(ii)	1	0.014	0.052	0.087	0.145	0.180
(iii)	1	0.013	0.050	0.083	0.139	0.171
(iv)	1	0.014	0.051	0.084	0.142	0.173
(vii)	6	0.018	0.056	0.088	0.132	0.157
(viii)	6	0.014	0.051	0.084	0.134	0.158
(x)	6	0.013	0.045	0.081	0.129	0.162

D) Winter

Model	Region(s)	1-step	2-step	3-step	4-step	5-step
(i)	1	0.021	0.051	0.088	0.119	0.159
(ii)	1	0.021	0.049	0.086	0.114	0.154
(iii)	1	0.021	0.047	0.081	0.108	0.146
(iv)	1	0.020	0.047	0.081	0.109	0.147
(vii)	6	0.023	0.050	0.090	0.122	0.157
(viii)	6	0.022	0.049	0.085	0.113	0.148
(x)	6	0.021	0.047	0.084	0.110	0.147

Table 3. Comparisons of MSEs for all seasons

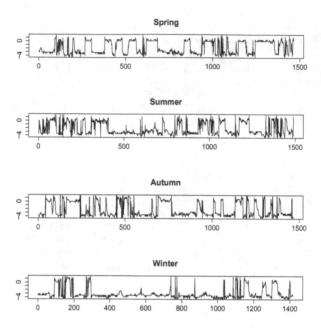

Figure 8. Changes in wind direction (rad.) for the four seasons (Matsumae)

5.2. Application of the model to estimate the impacts of wind flow on wave height

One of on the reasons why wave development phenomena are of interest is to improve understanding of how the direction of wind flow can impacts wave heights. To examine the applicability of the spatiotemporal model developed here, we estimate the wind flow above by applying the model. Figure 9 displays histograms showing wind direction for the four seasons, as observed in the data measured at six meteorological stations. Note that the horizontal axis corresponds to the wind direction shown at 16 azimuths, where 1, 5, 9, 13 corresponds to north, east, south and west, respectively. And Figure 10 shows histograms of the estimated values of s^* obtained using the proposed model (x). Here "MA", "OK", "ES", "MO", "HA" and "OH" correspond to the meteorological stations located at Matsumae, Okushiri, Esashi, Mori, Hakodate and Ohma, respectively.

We have examined whether or not it is possible to estimate wind flow that results in a large impact on wave height using the histograms shown in Figure 10. In spring, Figure 10 suggests that the meteorological stations at Hakodate, Matsumae and Esashi are capable of measuring the wind flow that significantly impacts wave motion at Matsumae-oki. In addition, Figure 9 shows that wind flows from the east and west are highly probable. For the case of westerly winds, based on Figure 1, wind motions measured over Okushiri, Esashi and Matsumae are thought to be highly correlated with wave-height changes at Matsumae-oki, where the open

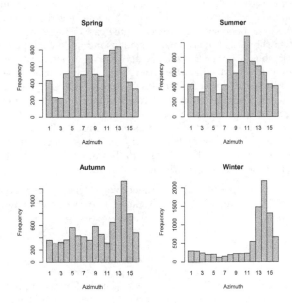

Figure 9. Histograms of wind directions for the four seasons

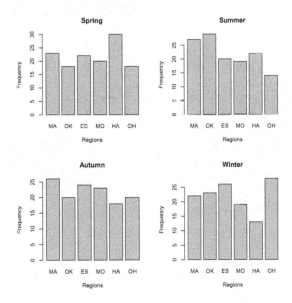

Figure 10. Histograms of estimated values for s^* for the four seasons

sea lies directly to the west of the towns above. In contrast, when the wind blows from the east, the wind motion over Hakodate, Ohma and Matsumae is thought to correlate with the wave height where the open sea lies directly to the east of these towns. Therefore, the evaluation of the data shows that the estimated stations in Figure 10, Hakodate, Matsumae and Esashi, can be classified with the five stations above. In summer, the dominant winds are southwesterlies (Figure 9). In such winds, the wind motions over Okushiri, Esashi, Matsumae and Hakodate are expected to correlate with the wave motion over Matsumae-oki, because there is open sea with sufficient fetch to the southwest of that location. In contrast, Figure 10 suggests that Matsumae, Okushiri and Hakodate are estimated as s^*, which is consistent with the towns above. In winter, strong seasonal winds are mainly northwesterlies. In this case, Ohma, Matsumae, Okushiri and Esashi are expected to have the best correlations with the wave motion for similar reasons as before. Figure 10 shows that Ohma, Esashi and Okushiri are estimated as s^*, and can be classified with the towns above.

Based on the considerations presented above, meteorological station measurements of wind flow, presented in the histogram of s^*, are shown to be effective for estimating wave-height changes at Matsumae-oki.

6. Conclusion

In this chapter, we have developed a statistical spatiotemporal model for forecasting wave-height changes and then applied it to the wave-height forecasting problem based on spatiotemporal wind motions measured at multiple AMeDAS meteorological stations. The results of the forecasting experiments have shown that the spatiotemporal model, that takes wind speed and wind direction into account, can improve forecasting accuracy when general time series models are used.

The spatiotemporal model presented in this chapter assumes that changes in wind direction follow the von Mises process. It may be possible, however, to further improve the forecasting accuracy by considering a stochastic process that enables a more flexible expression of changes in direction. The model improvements, including the consideration of directional processes, are expected to contribute to a deeper understanding of the transitional phenomenon that link wind motion and wave development, as well as the spatiotemporal processes involved with wind motion.

Author details

Tsukasa Hokimoto

Graduate School of Mathematical Sciences, The University of Tokyo, Japan

References

[1] Athanassoulis, G.A., Stefanakos, C.N. (1995). A nonstationary stochastic model for long-term time series of significant wave height, *Journal of Geophysical Research*, 100(C8), 16149-16162.

[2] Box, G.E.P., Jenkins, G.M. (1976). *Time Series Analysis, Forecasting and Control* (revised edition), Holden-Day, San Francisco.

[3] Breckling, J. (1989). *The Analysis of Directional Time Series: Applications to Wind Speed and Direction*, Lecture Notes in Statistics, 61, Springer-Verlag, Berlin.

[4] Brockwell, P.J., Davis, R.A. (1996). *Introduction to Time Series and Forecasting*, Springer-Verlag, New York.

[5] Erdem, E., Shi, J. (2011). ARMA based approaches for forecasting the tuple of wind speed and direction, *Applied Energy*, vol. 88, issue 4, 1405-1414.

[6] Guedes Soares, C., Ferreira, A.M. (1996). Representation of non-stationary time series of significant wave height with autoregressive models, *Probabilistic Engineering Mechanics*, 11, 139-148.

[7] Hokimoto, T., Shimizu, K. (2008). An angular-linear time series model for waveheight prediction, *Annals of the Institute of Statistical Mathematics*, 60, 781-800.

[8] Hokimoto, T. (2012). *Prediction of wave height based on the monitoring of surface wind*, Ocenanography, Chapter 8, 169-188, InTech, Rijeka, Croatia.

[9] Johnson, R.A., Wehrly, T.E. (1978). *Some angular-linear distributions and related regression models*, Journal of the American Statistical Association, 73, 602-606.

[10] Liu, H., Erdem, E., Shi, J. (2011). Comprehensive evaluation of ARMA-GARCH(-M) approaches for modeling the mean and volatility of wind speed, *Applied Energy*, 88, 724-732

[11] Philippopoulos, K., Deligiorgi, D. (2009). Statistical simulation of wind speed in Athens, Greece based on Weibull and ARMA models, *International Journal of energy and environment*, Issue 4, Volume 3, 151-158

[12] Pierson, W.J., Neumann, G, and James, R.W. (1960). *Practical Methods for Observing and Forecasting Ocean Waves by Means of Wave Spectra and Statistics*, U.S. Navy Hydrographic Office; Reprint edition.

[13] Scotto, M.G., Guedes Soares, C. (2000). Modelling the long-term time series of significant wave height with non-linear threshold models, *Coastal Engineering*, 40, 313-327.

[14] Scheffner, N.W., Borgman, L.E. (1992). Stochastic time-series representation of wave data, *Journal of Waterway, Port, Coastal and Ocean Engineering*, 118 (4), 337-351.

[15] Spanos, P.D. (1983). ARMA algorithms for ocean wave modelling, *Journal of Energy Resources Technology*, 105, 300-309.

[16] Sverdrup, H.U., and Munk, W.H. (1947). *Wind Sea and Swell: Theory of Relation for Forecasting*, U.S. Navy Hydrographic Office, Washington, D.C., No. 601.

[17] Tol, R.S.J. (1997). Autoregressive conditional heteroscedasticity in daily wind speed measurements, *Theoretical and Applied Climatology*, 56, 113-122.

Chapter 3

Challenges and New Advances in Ocean Color Remote Sensing of Coastal Waters

Hubert Loisel, Vincent Vantrepotte,
Cédric Jamet and Dinh Ngoc Dat

Additional information is available at the end of the chapter

1. Introduction

Knowing that coastal areas concentrate about 60% of the world's population (within 100 km from the coast), that 75-90% of the global sink of suspended river load takes place in coastal waters in which about 15% of the primary production occurs, the ecological, societal and economical value of these areas are obvious (fish resources, aquaculture, water quality information, recreation areas management, global carbon budget, etc). In that context, precise assessment of suspended particulate matter (SPM) concentrations and of the phenomena controlling its temporal variability is a key objective for many research fields in coastal areas. SPM which encompasses organic (living and non-living) and inorganic matter controls the penetration of light into the water and brings new nutrients into the system, both key parameters influencing phytoplankton primary production. Concentrations and availability of SPM are also known to control rates of food intake, growth and reproduction for various filter feeder organisms. Phytoplankton is highly sensitive to environmental perturbations (such as nutrient inputs, light, and turbulence). The abundance, biomass and dynamics of phytoplankton in coastal areas therefore reflect the prevailing environmental conditions and represent key parameters for assessing information on the ecological conditions, as well as on the coastal water quality. Because phytoplankton is highly sensitive to environmental perturbations [1], its distribution patterns and temporal variability represent good indicators of the ecological conditions of a defined region [2, 3]. Coastal waters also host complex ecosystems and represent important fishery areas that support industry and provide livelihood to coastal settlements. The food chain in the coastal ocean is generally short (especially in upwelling systems, having as low as three trophic levels) whereas the open ocean food web presents up to six trophic levels [4]. As a result, when compared to the open ocean, a relative lower fraction of the primary production gets respired in the coastal ocean while a higher fraction reaches the uppermost trophic level (fish) [5] or is exported to adjacent areas (coastal or open sea).

The potential large fluxes of carbon in the coastal ocean underscore its significance to the global carbon cycle (around 14% of total global ocean production, along with 80–90% of new production). However, the extreme heterogeneity of coastal environments in terms of ecosystems structure and functioning, and the current large uncertainties remaining at global scale on the latter features makes difficult general assessment of net autotrophic or net heterotrophic character of these marine waters. The results of the Shelf Edge Exchange Processes Study raised the question: "Do continental shelves export organic matter?". This question still remains without any definitive answer. Bauer and Druffel [6] suggested that dissolved organic carbon (DOC) and particulate organic carbon (POC) inputs from ocean margins to the open ocean interior may be greater than inputs of recently produced organic matter derived from the open surface ocean by more than one order of magnitude. While particles from terrestrial origin are primarily deposited in the coastal region, DOC is considered as the main path for transporting terrestrial and phytoplankton derived carbon into the deep ocean [7]. The accurate assessment of the temporal variability of SPM, POC, DOC, and Chl in coastal areas over long time periods and along with physical forcing parameters may therefore provide valuable insights for improving our knowledge on the biogeochemical cycle prevailing in coastal ecosystems.

Due to the high variability of the physical and biogeochemical processes occurring in coastal areas, traditional approaches based on oceanographic cruises and *in situ* time series, although essential, are very time-consuming, expensive and sometimes uncertain to yield meaningful results on the studied phenomena, especially at a large synoptic scale. In this context, remote sensing of biological and physical parameters is a very powerful tool for performing large scale studies. Satellite data are not as accurate as *in situ* measurements and are limited to the surface layer. However, the latter limitations are largely compensated by the spatial and temporal coverage offered by the satellite observations. *In situ* data remain obviously necessary to validate the satellite products, in terms of absolute value, but also in terms of temporal variability in areas where long *in situ* time series are available. After a short introduction of the main and critical aspects of ocean color radiometry, the specific issues relative to the satellite observation of ocean color in complex natural systems such as coastal marine waters will be specifically addressed. Illustrations of current algorithms development will be then provided as well as examples of the different end-users products currently available from ocean color remote sensing in coastal waters. At last, the new challenges and concepts allowing for a better observation of coastal ecosystems from satellite ocean color observation will be discussed.

2. General concepts of the ocean color radiometry

The interaction of light field within the visible part of the spectrum (i.e. 400-700 nm) with the different optically significant constituents of sea water (water molecules, salt, particulate and organic dissolved matters) modifies the color of the water. These spectral variations, which bring qualitative and quantitative information about the water constituents, can be recorded by a passive radiometric sensor onboard a satellite platform. However, the conversion of the radiometric signal measured by the sensor at the top of the atmosphere (TOA) to the end-user parameters is not straightforward. The contribution of the reflected photons at the air-sea interface as well as the contribution of the atmosphere should first be removed from the top of atmosphere measured signal to assess the water-leaving radiance, $L_w(\lambda)$, which is the only radiometric parameter encompassing useful information on the water masses composition (λ represents the wavelength of light in nanometer, nm). The removal of the atmospheric path radiance represents the most challenging part of the ocean color atmospheric correction procedure [8]. This signal, which includes Rayleigh and aerosols components, can contribute to up to 90% of the TOA signal depending on λ, the geometry of illumination and observation, the aerosol optical thickness, and the water leaving signal [9]. The importance of the latter task is particularly crucial when considering the level of accuracy required for being able to derive accurate estimates of the desired water components from $L_w(\lambda)$. As a matter of fact, the radiometric accuracy required for the Sea-viewing Wide Field-of-view Sensor (SeaWiFS) is, for instance, of 5% for the visible spectral domain for absolute radiance values, and of 2% for relative values (i.e. reflectances) [10].

The retrieval of the different inherent optical properties (IOPs) and biogeochemical components of the water from the water-leaving radiance spectral values are performed through bio-optical algorithms. Considering that about 90% of the $L_w(\lambda)$ signal originates from the upper layer of the water column (the so-called first attenuation layer), the various biogeochemical or optical variables which can be retrieved from $L_w(\lambda)$ are assumed to be weighted averaged parameters within this upper layer [11]. The thickness of this oceanic layer in the visible part of the spectrum typically varies from less than one meter (as in turbid waters, or in the red part of the spectrum) to about 60 meters (as for oligotrophic waters in the green), depending the amount of optically significant constituents within the water mass and the measured light wavelength [12]. Since the proof-of-concept Coastal Zone Color Sensor (CZCS) mission in 1978, satellite-derived ocean color has been routinely interpreted, with increasing accuracy, in terms of upper-ocean chlorophyll concentration, Chl [13]. The great availability of ocean color satellite data, as well as the need of complementary product (additionally to Chl) for validating global ocean biogeochemical models have greatly stimulated the rapid development of numerous inverse methods over the last decade for retrieving a large variety of information (Figure 1). For instance, empirical [14] and semi-analytical methods [15, 16] have been developed for specifically assessing the particulate organic carbon (POC) over the global ocean. In the same way, the various inherent optical properties are now routinely retrieved from space [17]. Among these different IOPs one may cite the coefficients of absorption by phytoplankton (a_{phy}) and colored detrital matter (a_{cdm}), as well as the particulate backscattering coefficient (b_{bp}). The spectral shape of b_{bp} [18] and a_{cdom} [19], which, respectively, provide relevant information on the particulate and dissolved pools, are also currently accessible over open ocean waters. Note that the retrieval of the bio-optical parameters can also be performed simultaneously to the atmospheric corrections using, for instance, neural network approaches [20]. Additionally, information about phytoplankton community composition [21-23] and size distribution of marine particles [24], essential for a better understanding of the oceanic carbon cycle, are also available from satellite imagery. With the ability to measure the Sun-stimulated phytoplankton *Chl* fluorescence, information on the physiological state of algal populations could also be derived from space as illustrated from the Moderate Resolution Imaging Spectroradiometer (MODIS) and MEdium Resolution Imaging Spectrometer (MERIS) ocean color missions. Detailed description of the different atmospheric and bio-optical algorithms for the observation of ocean color from space can be found in the extensive literature provided in the International Ocean Colour Coordinating Group web site [25].

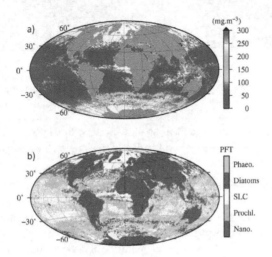

Figure 1. SeaWiFS global distribution of particulate organic carbon (a) and phytoplankton groups (b) in January 2001 as estimated by [15] and [26], respectively.

Thanks to its continuous recording of 13-years of data, the Sea-viewing Wide Field-of-view-Sensor ocean color mission (SeaWiFS) has allowed great scientific advances in our under-standing of the open ocean biological productivity [13]. In contrast, while ocean color observations allowed the achievement of numerous researches and new discoveries over open ocean waters, the algorithm development over coastal waters is still not mature enough to be applied routinely as it is currently performed for offshore waters, despites numerous recent advances. Some of the most important challenges remaining for developing accurate ocean color products in coastal waters are described in the following sections.

3. Challenges to overcome for ocean color radiometry in coastal areas

Remote sensing of ocean color in coastal areas is impacted by their intrinsic environmental features: vicinity of land over which photons can be reflected back to the sensor (the so-called adjacency effects), land inputs from rivers discharges and coastal washing, shallow waters which promote resuspension of the unconsolidated sediments as well as bottom reflected

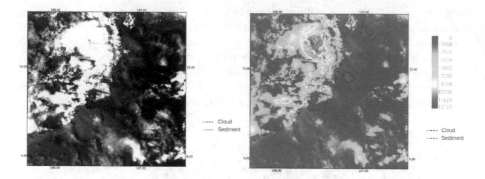

Figure 2. left) RGB composite of a daily MODIS image over the northen Vietnam coastal waters. (right) Top of atmosphere signal (in numerical counts). This figure shows that sediment dominated waters and clouds may have the same radiometric signature at the top of the atmosphere.

photons captured by the sensor in clear shallow waters, breaking waves which generate bubbles, etc. All these phenomena make difficult and challenging the assessment of bio-optical marine components from space by ocean color sensors.

The main problems to overcome in order to derive accurate observations of coastal areas from ocean colour remote sensing are associated with the presence of highly diffusing suspended mineral particles as well as high concentrations of particulate organic matter (phytoplankton, detritus) which may bias atmospheric corrections and impact bio-optical algorithms [17, 27]. Another issue largely encountered for the exploitation of ocean colour data in coastal areas is the presence of clouds. In presence of highly turbid waters, cloud-free pixels are sometimes erroneously classed as clouds, leading to a loss of data. At last, but not least, crucial and mandatory validation exercises are greatly complicated to perform due to the extreme spatial heterogeneity of these areas. These different issues will be addressed in the following sections.

3.1. Cloud masking

Ocean color retrieval by satellite borne sensors is in principle only possible for a clear (cloud free) atmosphere, and cloud-contaminated pixels have to be removed from the images before any ocean color processing takes place. The cloud masks developed in the frame of ocean color missions are based on the assumption that the marine reflectance in the near-infrared (NIR) is equal or close to zero. While open ocean waters can effectively be considered as black in the NIR, except in presence of highly scattered material (due for instance to offshore river plumes, or coccolithophore blooms), this is not the case for coastal waters. Indeed, the level of signal observed at the top of the atmosphere over clouds and coastal areas may be similar due to the presence of suspended sediments in the water surface (Figure 2). Cloud-free pixels are then sometimes classed as clouds and excluded from further processing, leading to a loss of data in these areas. Specific cloud masking algorithms have to then be developed over coastal waters.

3.2. Atmospheric corrections

a. General description of the atmospheric correction

The use of satellites to monitor the color of the ocean requires effective removal of the contribution of the atmosphere (due to absorption by gasses and aerosols, and scattering by air molecules and aerosols) to the total signal measured by the remote sensor at the top of the atmosphere, L_{toa}: the so called "atmospheric correction" process. As shown in Figure 3, the signal measured by the remote sensor is the sum of different components. The top-of-atmosphere radiance, L_{toa}, can be expressed as [8]:

$$L_{toa}(\lambda) = L_r(\lambda) + L_{ra}(\lambda) + L_a(\lambda) + t(\lambda)xL_{wc}(\lambda) + T(\lambda)xL_g(\lambda) + t_0(\lambda)xL_w(\lambda) \tag{1}$$

where L_r is the radiance due to the scattering of the air molecules (Rayleigh scattering), L_{ra} is the multiple interaction term between molecules and aerosols, L_a, the radiance due to the scattering by aerosols, L_{wc}, the radiance due to the whitecaps on the sea surface, and L_g the specular reflection of direct sunlight off the sea surface. $t(\lambda)$ and $t_0(\lambda)$ are the diffuse transmittances of the atmosphere from the sun to the surface and from the surface to the sensor, respectively, $T(\lambda)$ is the direct transmittance from the surface to the sensor, and $L_w(\lambda)$ the water-leaving radiance [28].

The terms L_{wc}, L_g, and L_r can be determined during a pre-processing. So the atmospheric correction algorithm needs to solve the following equation:

$$L_{cor}(\lambda) = L_{toa} - L_r(\lambda) - t(\lambda)x\rho_{wc}(\lambda) - T(\lambda)x\rho_g(\lambda) = L_a(\lambda) + L_{ra}(\lambda) + t(\lambda)xL_w(\lambda) = L_A(\lambda) + t_0(\lambda)xL_w(\lambda) \tag{2}$$

The goal of the atmospheric correction process is to accurately determine $L_w(\lambda)$ from the spectral measurements of $L_{cor}(\lambda)$. For that purpose $L_A(\lambda)$ has to be quantified.

The classic methods for removing the contribution of the atmosphere to the total measured signal exploit the high absorption by seawater in the red and near-infrared (NIR) spectral regions. In open seawater, i.e. where generally chlorophyll-a concentration and related pigments and co-varying materials (like detritus) determine the optical properties of the ocean, seawater can be considered to absorb all light in the NIR so that the signal observed by the satellite sensor in this spectral domain is assumed to be entirely due to the atmospheric path radiance (L_A) and sea surface reflectance [8]. This is not always the case when considering turbid waters (generally coastal optically complex waters). In these waters phytoplankton pigment and detritus, as well as suspended sediment, contribute to NIR backscatter. The resulting NIR water-leaving radiances introduce two sources of error into the standard procedure to remove the aerosols. First, the total aerosol reflectance is overestimated as some of the total radiance (L_{toa}) at two NIR bands (λ_1 and λ_2) comes from the seawater. Second, as the absorption and scattering properties of seawater change from λ_1 to λ_2, the selection of the appropriate atmospheric model is affected, causing an error in the extrapolation of L_A toward shorter wavelengths. As a result, L_A is overestimated at all bands with increasing values at

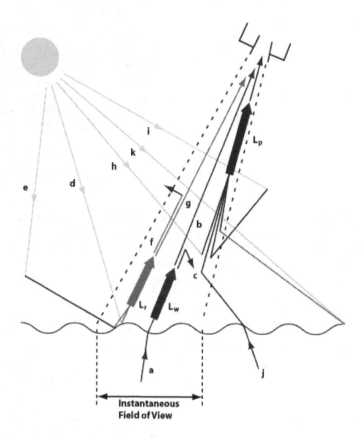

Figure 3. Illustration of several different light pathways in the atmosphere: a) The light path of the water-leaving radiance, b) the attenuation of the water-leaving radiance, c) scattering of the water-leaving radiance out of the sensor's field of view (FOV), d) sun glint (reflection from the water surface), e) sky glint (scattered light reflecting from the surface), f) scattering of reflected light out of the sensor's FOV, g) reflected light is also attenuated towards the sensor, h) scattered light from the sun which is directed toward the sensor, i) light which has already been scattered by the atmosphere which is then scattered toward the sensor, j) water-leaving radiance originating out of the sensor FOV, but scattered toward the sensor, k) surface reflection out of the sensor FOV which is then scattered toward the sensor. L_w Total water-leaving radiance. L_r Radiance above the sea surface due to all surface reflection effects within the FOV. L_p Atmospheric path radiance. This Figure is adapted from [29].

shorter wavelengths, even possibly leading to negative water-leaving radiances in the blue bands [30].

The difficulty of the atmospheric correction is that the atmosphere contributes to 80-90% of the total top-of-atmosphere signal at the blue-green wavelengths (400-550 nm) and the atmospheric path radiance L_A significantly varies and cannot be easily approximated. Moreover, the assumption of having a non-zero L_w in the NIR bands is not valid for turbid waters (Figure 4). To solve this problem, several specific atmospheric correction algorithms have been developed in the past decade for the major past and current ocean color remote sensors [31-46].

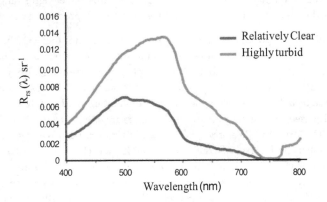

Figure 4. Spectral remote-sensing reflectance measured in relatively clear (red line), and turbid (blue line) waters of the French Guyana.

2. Description of existing algorithms

The algorithms previously cited can be grouped into four different categories:

1. Assignment of hypothesis on the NIR aerosols or water contributions

For these algorithms, assumptions are made on the aerosol models and/or the bio-optical models. Hu et al. [35] estimate the aerosol model in clear waters on the studied image and extrapolate this information over the coastal waters of the image. Ruddick et al. [40] impose a constant value of the ratio of L_A and L_w in the NIR. The aerosol constant NIR ratio value is determined for each image of interest taking into account only the clear waters as in [35]. The constant $L_w(NIR)$ ratio is defined from radiative transfer modeling and is assumed to be constant for any location for a given satellite sensor. This ratio is valid for low and moderate turbid waters. Using other wavelengths (SWIR, see next subsection), Wang et al. [46] developed a similar scheme than the one proposed by [35] for the Lake Taihu (China). The aerosol properties are obtained at the pixel-by-pixel level. The derived aerosol model is averaged in the middle of the lake, where the waters are not turbid. Using this mean aerosol model, the SWIR atmospheric correction algorithm use a single SWIR band and can be carried out for the entire lake for estimating L_w. These assumptions allow to use the scheme defined in [8] for the dark pixel hypothesis. These constant ratios only allow to determine $L_A(NIR)$ and to remove this contribution to L_{cor}.

2. Use of the shortwave infrared bands

For moderate and very turbid waters, it is possible to use the shortwave infrared bands (from the MODIS-AQUA sensor). The principle of this type of algorithm is very similar than the one of Gordon and Wang [8], except that the aerosols models are determined for wavelengths between 1200 and 2100 nm instead of NIR wavelengths. For this wavelengths range, L_w can be considered negligible and the dark pixel atmospheric correction processing can be applied.

The difficulty with SWIR algorithms is the low signal to noise ratio for the SWIR bands of the MODIS-AQUA sensor. The algorithm shows limitations for low and moderate turbid waters, and therefore needs to be coupled with a NIR atmospheric correction algorithm [47]. More specifically, the latter authors have proposed to use the standard NIR-based atmospheric correction unless a turbidity index [48] exceeds a predefined threshold and, thereby, triggers the use of a SWIR-based correction where two SWIR bands are used instead of two NIR bands. The relevant MODIS-Aqua SWIR band pair is 1240 and 2130 nm. Shi and Wang [48] concluded that these bands satisfy the black pixel assumption in moderately (1240 nm) to extremely (2130 nm) turbid waters. This combined "NIR–SWIR" atmospheric correction approach has been evaluated extensively in several geographic locations [34, 47, 48] and validated against an independent, globally-distributed in situ data set [45]. Moreover, the difficulty of using the SWIR bands relies on the selection of the aerosol model. Analysis of the Ångström exponent retrievals suggests that the SWIR approach cannot retrieve the correct aerosol spectral dependence when the atmospheric path radiance is low. The latter feature tends to produce spectrally flat models, thus underestimating the aerosol reflectance in the shorter wavelengths. In fact, at low signals the SWIR approach often predicts aerosol spectral dependence that is beyond the range of the current aerosol models, as it would be associated with unrealistically large aerosol size distributions. Conversely, the SWIR approach tends to overestimate aerosol reflectance at higher aerosol loads, thus resulting in negative $L_w(\lambda)$ [49].

3. Correction/modelling of the non-negligible L_w(NIR)

This type of algorithm aims to correct the non-negligible L_w(NIR) signal using iterative methods coupled with a bio-optical algorithm [30, 31, 38, 43, 46]. These algorithms need to define a first value of L_w(NIR) which can be directly fixed, or calculated from a first guess of *Chl* or IOPs. This can be done by using either default values of the parameters or a dark pixel atmospheric correction procedure such as [8]. Then, a NIR water-leaving reflectance model, developed from *in situ* measurements, is used to calculate L_w(NIR). The problem encountered with these algorithms is that the empirical bio-optical models limit their applicability to waters that are similar to those over which these empirical models were developed.

4. Coupled ocean-atmosphere inversion

As the ocean and the atmosphere cannot be decoupled in turbid waters (L_w(NIR) $\neq$ 0), an appropriate solution for developing atmospheric correction scheme would be to couple the two systems and to inverse them together. In practice, this can be done using either artificial neural networks (NN) [41] or optimization techniques [20, 31, 36, 41, 50-54].

The more common type of NN is the Multi-Layer Perceptron (MLP) [55]. Thanks to this association of elementary tasks, an MLP is able to solve complex inverse problems. The specificity of an MLP depends on the topology of the neurons (number of layers, numbers of neurons on each layer) and on the connection weights w_{ij} from a neuron j of a layer to the neuron i of the next layer. The w_{ij} weight values are computed through a training phase, using a training dataset. Once the training phase is finished, the MLP will only do algebraic operations, which leads to faster computations, which is very convenient for satellite applications. When using a neural network for the atmospheric correction phase, the top-of-atmosphere radiance is usually directly inverted for estimating the water-leaving radiance in the visible as well as the aerosol optical properties. Another option is to use optimization techniques. The principle of those algorithms is based on the iterative minimization of a dedicated cost function by adjusting relevant atmospheric (aerosol optical thickness, and Ångström coefficient) and oceanic (Chlorophyll concentration, IOPs) parameters which are the control parameters [20, 56]. There are some difficulties that one encounters when using this type of techniques: the parameters initialization [57], definition/use of a direct model, calculation of the adjoint code and the method for the parameters adjustment [56].

2. Are these algorithms accurate enough?

As shown previously, there are several methods to estimate L_w from L_{toa} in complex coastal waters. Few round-robin comparisons have been made to evaluate different atmospheric correction algorithms from *in situ* data [58, 59]. These studies showed that the different algorithms highly over-estimated L_w in the blue and red and were quite accurate for the intermediate wavelengths providing similar estimates of L_w (Figure 5). But the retrieval accuracy in the blue is still too low, with negative L_w values in some cases, which prevent to use these bands for bio-optical applications.

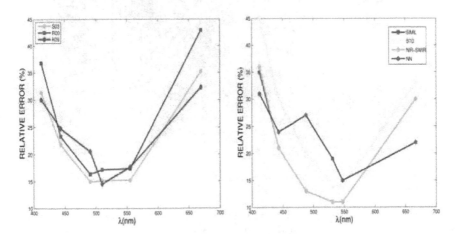

Figure 5. Evaluation of three SeaWiFS (left) and four MODIS-AQUA (right) atmospheric correction algorithms from in-situ data in complex coastal waters (Figures from [58] and [59], respectively).

3. Other issues

Besides the presence of water leaving radiance signal in the near infrared, other issues can be encountered during the atmospheric correction processing. The three major problems are the bottom albedo for shallow waters, the adjacency effects and the presence of absorbing aerosols.

The bottom effect corresponds to the light reflected off the bottom of a water body, providing the water is sufficiently shallow and clear. The influence of the seafloor on the colour of water depends on the depth of the water body, the clarity of the water masses, the type of optical substances present in the water, as well as on the type of substrate composing the seafloor. The bottom may be rocky or sandy, and may be covered, partially or fully, by a variety of benthic organisms (e.g., algae, mollusks). All of these factors will influence the manner in which bottom effects will act on the colour of the water, as seen by a remote sensor [27]. Figure 6 shows the particulate backscattering coefficient estimated at 490 nm using [60]. The area in light pink and white correspond to negative R_{rs} (and then b_{bp}) values induced by this bottom effect. If this effect is not corrected before or during the atmospheric correction, the estimation of the R_{rs} (b_{bp}) will be biased and will lead often to obtain negative values. This is particularly true when studying clear shallow waters such as lagoons or the Coral Reefs Bareer in the east coast of Australia [61-63].

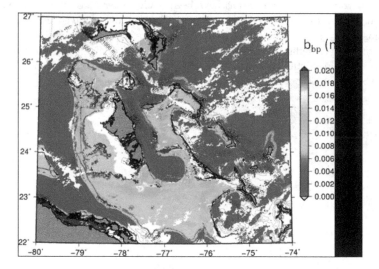

Figure 6. MODIS map showing the distribution of the backscattering coefficient of particulate matter (b_{bp}, m^{-1}) estimated over shallow waters of the Bahamas using [60].

The so called "adjacency effect" refers to the process by which a photon, reflected from a surface adjacent to a targeted pixel, is scattered by the atmosphere between the sensor and the target, blurring the sharp boundary between the land and the coastal water [64-66]. In coastal areas, a fraction of the light reflected by the land can reach the sensor. Modeling the reduction of image contrast when the atmospheric turbidity increases requires large amounts of computing time. The adjacency effects impart a significant bias to the model for retrieval of aerosol values because of the high contrast between land and ocean in the near infrared spectral region. First, the aerosol optical thickness is overestimated by classic algorithm because of the increase of the atmospheric signal. Then the wavelength dependence is modified, leading to a wrong extrapolation of the aerosol optical thickness in the blue-green region. The variation of the aerosol optical thickness between the near infrared and the red on a transect from the coast can be a good indicator of contamination by adjacency effects of the aerosol product over land and an indicator of this effect on satellite imagery [66].

The third major problem is the presence of absorbing aerosols from urban and desert origin in coastal regions. The aerosol models used in the classic atmospheric correction are all non- or weakly absorbing [8, 9, 67]. Absorbing aerosols have a lower aerosol radiance at the shorter wavelengths than any of the models used in the current atmospheric correction process. The presence of absorbing aerosols would, thus, have a similar effect to the aforementioned incorrect model selection, though likely to a more significant degree in the blue portion of the spectrum. The presence of such strongly-absorbing aerosol can only be inferred in the visible, where multiple scattering is high. In such situations, one can no longer derive water properties by a two-step process — atmospheric correction followed by a bio-optical algorithm to estimate water properties. To solve this issue, only few methods have been currently proposed [20, 32, 51, 53, 68, 69]. A promising way to deal with the presence of absorbing aerosols would be to use spectral optimization method which allows the atmospheric and oceanic properties to be retrieved simultaneously.

3.3. Bio-optical algorithms

Satellite remote sensing of ocean colour is a very powerful tool for the management of resources and activities of continental shelf waters. Besides estimates of *Chl*, which represents the historical parameter investigated from optical remote sensing technique, more recent developments have allowed the retrieval of a variety of bio-optical (e.g. phytoplankton and detrital matter absorption or particulate backscattering coefficients) and biogeochemical (e.g. *POC*, phytoplankton size distribution and community composition) parameters with now a satisfying accuracy at global scale [70]. The latter developments have been supported by the improvement of optical sensors, from the first (CZCS launched in 1978) to the most recent ones (e.g. SeaWiFS, MODIS or MERIS), both regarding the radiometric data quality and spatial resolution.

An accurate assessment in the coastal marine domain of the various optical and biogeochemical parameters previously cited and now available for oceanic waters still represents an important challenge since it would provide relevant and innovative insights on the dynamics and functioning of these complex and highly diverse ecosystems. The development of innovative bio-optical algorithms is for instance cruelly needed for precisely identifying the occurrence of specific phytoplankton species. As a matter of fact, a precise monitoring of harmful algal bloom (HAB) events which strongly impact the functioning of the concerned coastal ecosystems and which dynamics are currently relatively poorly constrained, represents a crucial field of investigation. Various algorithms have, in that sense been, proposed in the recent years for computing indices specifically dedicated to the identification of red tides events [71-73]. Among the various important challenges to overcome in coastal waters, one particularly crucial concerns the estimation of dissolved and particulate organic carbon stocks and dynamics which would strongly condition our ability to quantify the impact of coastal ecosystems domain on the global carbon cycle. The assessment of POC and DOC concentrations in coastal waters however still represents an open field of investigation even though some empirical regional algorithms have been already proposed for estimating DOC content from CDOM absorption coefficient retrieval in some coastal regions [74-76].

The assessment of the latter bio-optical and biogeochemical parameters from space depends on the relevance of the bio-optical algorithm required to infer marine water IOPs or biogeochemical products from satellite $L_w(\lambda)$ measurements. This key step for interpreting the satellite signal in these marine regions is complicated by various specific issues. Indeed, coastal waters are highly dynamic systems at both temporal and spatial scales being subjected to a variety of physical (tides, current, fronts, turbulence...) and environmental forcings (interactions with terrestrial ecosystems, phytoplankton blooms, re-suspension...). This high hydro-biological variability of these waters logically induces a strong optical complexity (Figure 7) making more difficult the interpretation of the 'color' in these environments. Specifically, unlike in case 1 waters, substances significantly active from an optical point of view (i.e. phytoplankton, non-algal particles and colored dissolved material) usually vary independently in time and space and have specific optical properties which may vary over wide ranges [77-79]. These specificities of coastal waters create different issues. First, a strong dispersion around the average general laws generally used in open ocean waters for linking the radiometric measurements to bio-optical (absorption and backscattering coefficients) or biogeochemical parameters (e.g. chlorophyll a). The latter feature prevents the use of Case 1-like general relationships and emphasizes the crucial need to develop specific inversion schemes allowing to take into account for the peculiar optical characteristics of the coastal marine domain [27, 80, 81]. Second, standard algorithms used for estimating ocean colour such as Chla concentration in oceanic waters are based on reflectance ratios in the blue and green spectral domain [27]. In coastal waters, the significant contribution of dissolved and particulate non-algal material to the optical budget at these latter wavelengths complicates our ability to detect phytoplankton pigment optical signal and often induces a strong failure in the accuracy of these classical formulations (Figure 8).

Accurate assessment of the different in water bio-optical components from ocean colour measurements in coastal areas are therefore largely controlled by: (i) our ability to understand and to account for the origin of the variability in the radiometric signal and (ii) the realism of the parameterizations used between the inherent optical properties (IOPs) and the biogeochemical component (BC). Numerous challenges still remain [17, 27] in that sense, however some progresses have been recently performed by developing inversion approaches specifically dedicated to face the various issues encountered in coastal waters (see sections 3.3 and 3.4).

Figure 7. Left: MODIS true color image of a red tide event occurring in the coastal water of Texas. Right: SeaWiFS and MODIS maps showing quantitative estimates of phytoplankton Chla associated with red tides events in the coastal zone of Florida (http://earthobservatory.nasa.gov).

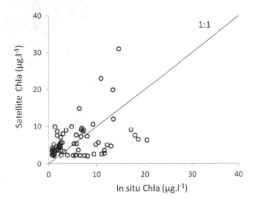

Figure 8. SeaWiFs (OC4V4 algorithm) matchups of *Chla* performed in coastal areas of the English Channel and southern north Sea (Loisel unpublished data).

3.4. Validation

Validation of ocean color products (i.e. $L_w(\lambda)$, IOPs, and biogeochemical parameters) should theoretically be performed from in situ measurements acquired simultaneously to the satellite overpass over the same location. However, these ideal conditions are rarely encountered and specific protocols should be applied [82, 83]. For instance, it is in practice recommended to use a 2-hour time window applied to the satellite overpass time at the measurement site to select satellite imagery. Second, the match-up procedure extracts a 3-by-3 pixel satellite image box with the middle pixel closest to the measurement site. The mean value of the desired parameter over the box is calculated for each image. Lastly, an atmospheric spatial uniformity criterion is applied, based on a prescribed coefficient of spatial variation, defined as the ratio of the standard deviation to the mean pixel value within the selected image box. This match-up exercise is made more complicated for coastal waters where typical high frequency physical and biological processes cause high temporal variability and strong spatial heterogeneity (Figure 9). In situ measurements are generally performed from oceanographic vessel or instrumented fixed buoy from which the spatial heterogeneity of the satellite pixel can not be necessarily taken into account. Typical pixel size of ocean color missions is about 1x1 km² at nadir, even if higher spatial resolution data (about 250x250 m) are now accessible for some specific wavelengths (i.e. MODIS), or over the whole spectrum but only on request, that is not routinely (i.e. MERIS-FR). Thanks to the very recent development of autonomous underwater vehicles such as gliders, equipped with optical and radiometric sensors, the spatial heterogeneity of a given pixel could now be sampled *in situ*. However, much more researches are needed to reach the level of accuracy needed for ocean color validation, from measurements collected using these new platforms.

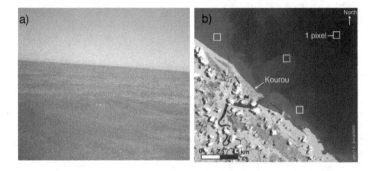

Figure 9. a) Picture of turbid patches found in the French Guyana coastal waters taken during a scientific mission in October 2009 [84]. (b) High resolution SPOT image showing the wide optical dynamics of the coastal waters of French Guiana. The white square Figures a 1x1 km pixel typically considered for ocean color remote sensing application.

4. New and future developments of ocean color remote sensing of coastal areas

4.1. Cloud masking

Prior to any application of atmospheric correction and bio-optical algorithms a cloud masking algorithm should be applied over the satellite observed area. For that purpose, different algorithms which differ by their inherent assumptions and the wavelengths used have been developed [85-89]. The standard algorithms to process the image acquired by SeaWiFS [85] and MODIS [89] use a constant threshold value at 865 nm (SeaWiFS) and 869 nm (MODIS) over which the pixel is considered as cloud. Wang and Shi [86] present a more sophisticated algorithm based on the fact that water is almost totally black in the shortwave infrared (SWIR) wavelengths. The latter authors have also proposed an alternative algorithm for ocean color sensors which do not allow for measurements in the SWIR bands (such as SeaWiFS and MERIS). Basically, this algorithm is based on the assumption that the spectral variability in the NIR is lower for clouds than for water. Based on the observation that water is generally spatially more homogeneous than clouds, the cloud mask for the POLarization and Directionality of the Earth's Reflectances (POLDER) sensor uses a threshold on the spatial variability in 865 nm [87]. To be able to automatically process and exploit the long time series of SeaWiFS ocean color images over coastal areas, a recent algorithm has been developed for the SeaWiFS sensor [88]. This algorithm is mainly based on the spectral differences of the Rayleigh free TOA signal between cloud and turbid waters. Performance of different algorithms for the very turbid waters of the Amazon delta is provided in Figure 10.

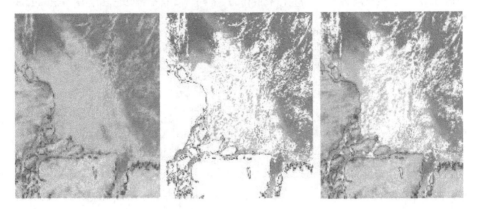

Figure 10. A color composite SeaWiFS image recorded over the Amazon delta area with no cloud mask (left), and after the application of the standard SeaWiFS NIR threshold (middle), and Nordkvist et al. [88] algorithm (right). The gain of clear water pixels is clearly seen when the latter algorithm is used instead of the standard SeaWiFS algorithm.

4.2. Atmospheric correction

As shown in Figure 5, the accuracy of the existing atmospheric correction algorithms is still not enough satisfying, especially when considering the blue and red spectral bands. It is therefore still necessary to develop innovative schemes to decrease the errors at these bands. As mentioned previously, one of the most promising algorithm development suggests the use of the shortwave infra-red bands (SWIR) for estimating the atmospheric path radiance [44, 47]. However, because of the low signal-to-noise ratio of the MODIS-AQUA SWIR bands, which induces uncertainties in the extrapolation of the aerosol models from the SWIR to the NIR and visible, and also due to the fact that none of the future ocean color sensors (OLCI, GOCI-II, OCAPI, S-GLI) will possess the SWIR bands, it is necessary to find other ways to improve the atmospheric correction procedure. Another promising option would be to consider the shorter wavelengths, i.e. in the ultraviolet bands [90]. For highly productive waters with high amount of colored dissolved organic matter, the water-leaving reflectance at 412 nm can be used to constrain the aerosol model retrieval considering that $L_w(412)$ is relatively low being negligible when compared to L_w in the visible and NIR bands. Considering the latter feature, UV bands could be used to determine the aerosol model. The major disadvantage of using the UV for the atmospheric correction is the assumption of a negligible L_w for these bands which are important for the retrieval of CDOM. This assumption needs to be further validated by increasing the amount of dedicated *in situ* data in the future.

Another way for improving atmospheric correction schemes in coastal waters would be to further develop the use of the formulations which consider a definition of the NIR bio-optical algorithms, by adding constrains on the relationships between L_w at several wavelengths. Several relationships already exist. For instance, Wang et al. [46] proposed a new atmospheric correction algorithms for the GOCI sensor using a relationship between $L_w(765)$ and $L_w(865)$. However, this algorithm needs in a preliminary step to calculate the diffuse attenuation coefficient, $K_d(490)$ for estimating $L_w(765)$ from the MODIS-AQUA sensor using the SWIR atmospheric correction algorithm. This step might add errors and complexity for developing such algorithms. This emphasizes also the need for developing more direct approaches. One solution would be to mix a new formulation of the NIR similarity spectrum approach [40, 91] such as defined in [59, 92] with the algorithm developed by [31] which has been demonstrated to be robust in low but also very turbid waters. Figure 11 presents evaluation of the similarity spectrum [40, 91] and formulation of Doron et al. [92] against in-situ data for moderately and highly turbid waters. We can see that the constant $R_{rs}(NIR)$ ratios is not valid for those waters, notably for highly turbid waters. These ratios could be integrated inside a current atmospheric correction algorithm for constraining the inversion. Another way is to find linear or polynomial relationships between two R_{rs} (Figure 12).

4.3. Bio-optical algorithms: The classification approach vs. regional approaches

Different approaches have been considered in the recent years for developing inversion algorithms in order to face the issue represented by the strong optical complexity and heterogeneity of coastal waters (see section 3.3). Two contrasted approaches based on a geographical or optical partition of the coastal waters will be considered here, both aiming to constrain the dispersion around bio-optical $R_{rs}(\lambda)$-IOPs-BC relationships and presenting their own advantages and requirements.

The first approach, which is the more usual one, consists in focusing on the range of optical variability specific to a defined coastal area by developing local or regional, usually empirical inversion algorithms [93-95]. This way of dealing with the optical complexity of coastal waters, although relatively convenient to implement (as soon as a reasonable amount of in situ measurements are available for a defined region) might present different limitations in its application. Indeed, while such regional algorithms may reduce the variability in the IOPs-BC relationships, they are highly dependent on the dataset used for their development. In other terms, the accuracy of regional relationships mostly depends on their ability to account for the

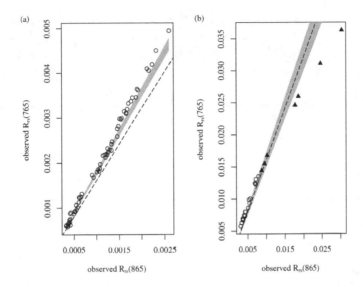

Figure 11. Reflectance ratio at 765 and 865 nm according to [91] (dashed line) and [92] (grey polygon) for (a) moderately and (a) highly turbid waters [59]. Circles and triangles are for data collected in the Eastern English Channel/North Sea and the French Guiana, respectively.

natural variability of the water masses optical properties occurring in a defined region. However, a relationship defined for a given area is likely to vary over time (i.e. at the daily scales, seasonal or inter-annual) since it depends strongly on the combined action of various environmental forcing (phytoplankton blooms, in suspension, leaching of the sides, riverine inputs and advection of the offshore waters....). Furthermore, considering the high dynamics of coastal ecosystems it might be presumably challenging to capture, even for a defined region, the many high-frequency processes affecting bio-optical relationships. Finally, in the scope of applications of remote sensing techniques in coastal waters over large spatial scales (i.e. global), the use of a regional approach (by definition non-exportable geographically) would consist in performing a collection of algorithms developed on a mosaic of coastal areas. This seems both difficult to implement in practice and would inevitably lead to ignore some coastal areas currently not covered by *ad hoc* in situ measurements.

Another way to deal with the optical heterogeneity of the coastal domain is to specifically consider the optical diversity of these waters within the algorithm development procedure. In practice, this alternative approach to regional methods is based on optical classifications, which aims at grouping waters with similar optical features and develops optically adapted algorithms for each water class. Such a definition of optical water types implicitly assumes that different coastal regions can present similar optical characteristics of the marine components

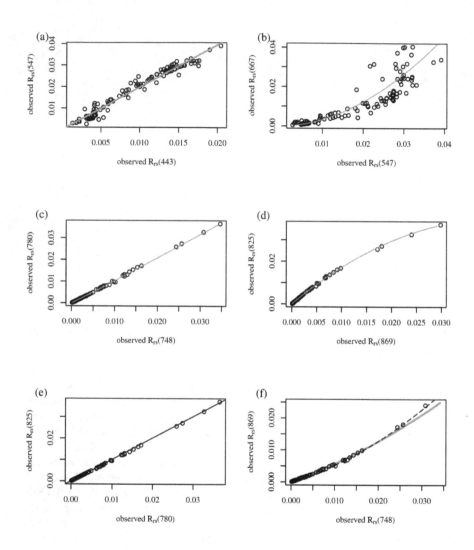

Figure 12. a) Linear relation between Rrs(λ) at 443 and 547 nm, b) Polynomial relation between Rrs(λ) at 547 and 667 nm, c) constant ratio between Rrs at 748 and 780 nm, d) Polynomial relation between Rrs(λ) at 820 and 869 nm, e) constant ratio between the modeled Rrs(λ) at 780 and 825 nm, f) polynomial relation between Rrs(λ) at 748 and 869 nm [59]. For this latter figure, the dotted line represents the polynomial regression line over the in-situ dataset. The grey thick line represents the polynomial relationship defined by [59].

which translate into similar reflectances, at least during some moments of the year. Classification-based approaches are therefore intrinsically independent of the location and the time, and thus present a stronger universal character being presumably more suitable for large scale applications than classical regional approaches. Previous studies have treated the partition of the marine waters into optical classes. They were based on sets of bio-optical parameters diversely including the following: the diffuse attenuation coefficient, the Secchi depth or inherent optical properties and biogeochemical parameters [12, 96-98]. Other studies, in particularly, at global scale, were dedicated to optical water type definition based on the information provided by the remote sensing reflectance spectra [99, 100]. These studies have emphasized the potential of using ocean typology based on ocean color radiometry for characterizing the uncertainties related to ocean color products [101] and improving the performance of the inversion procedure [102]. Furthermore, the advantage of the optical classification approach for providing relevant insights into ocean water masses optical dynamics and therefore studying its biogeochemical quality has also been illustrated for the open ocean waters [103]. Similar applications of optical classification dedicated to the coastal ocean are currently limited. Few studies have specifically focused on a characterization of the optical diversity of coastal water masses from in situ measurements [83, 105].

Satellite applications of coastal ocean optical typology for monitoring coastal water quality [106] or improving ocean color product inversion [107, 108] are also still relatively scarce. A recent study [109] performed from an in situ data set gathered in contrasted coastal waters (i.e. eastern English Channel, southern North Sea and French Guiana) has emphasized the applicability and the advantages of this approach (Figure 13). Specifically the main results of this study have emphasized that (i) the ubiquitous character of R_{rs} spectra optical shape at global scale (ii) the need of a reasonable amount of optical classes for describing coastal waters optical diversity (iii) the interest of the optical classification for dynamically monitoring the coastal waters masses quality and (iv) the potential for this approach for improving estimates of satellite products (with preliminary results on the SPM retrieval). The potential of the classification-based approach should be however explored more in detail through the estab-lishment of large data set coupling optical and biogeochemical measurements gathered in a large variety of coastal waters.

4.4. The need of other "tools"

a. Towards the use of other spectral domains

The past and present ocean color sensors are generally characterized by a common set of spectral channels. The blue (412, 443, 490 nm) and green (510, 550/560 nm) wavelengths are used for bio-optical algorithms purposes whereas the red (670) and near infrared (765, 865 nm) wavebands are used for atmospheric corrections. To these standard ocean color channels, some sensors also have the ability to measure the chlorophyll-a fluorescence peak using a band centered at about 676 nm in combination with two surrounding bands (around 665 and 746 nm) used for the baseline [110]. To overcome the challenges of ocean color remote sensing of coastal areas, the use of visible wavelengths other than the standard ones should also be examined. For instance, the maximum reflectance signal observed in coastal areas is, in many cases, observed around 590 nm, which is far from the 550/560 standard bands which can not adequately capture this maximum (Figure 14). Because of the generally low absorption

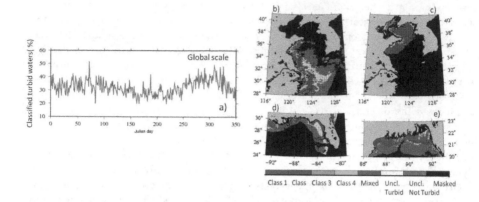

Figure 13. a) Annual evolution of the percentage of SeaWiFS turbid pixels [85] labeled as belonging to one of the four optical water types defined from our *in situ* data set at global scale. Illustrations of the optical water types distribution (b,c) in the China sea coastal waters, d) in the northern Gulf of Mexico and e) in the coastal waters influenced by the Gange-Brahmaputra rivers output. White pixels correspond to unclassified turbid waters, light and dark grey show non-turbid and masked regions, respectively (taken from [110]).

coefficient at this wavelength, it could also be used to improve the determination of the spectral slope of the particulate backscattering coefficient [18].

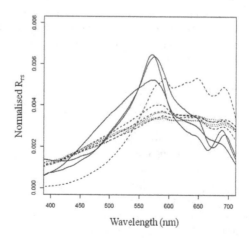

Figure 14. Example of normalized remote sensing spectra, R_{rs}, collected in various coastal environments (English Channel, North Sea, and French Guyana). The description of this data set can be found in [105]. The normalization has been performed by dividing raw R_{rs} spectrum by its integrated value in order to emphasize changes in the shape of the reflectance found in coastal waters.

The recent use of the ultraviolet (UV) bands in the frame of the development of ocean color algorithms over coastal waters has shown some promising results. For instance, the exponential increase of detritus and colored dissolved organic matter absorption coefficient from the long to the short wavelengths makes the UV domain an ideal candidate for atmospheric correction purposes in turbid waters [90]. This new method, which based on a low and relatively stable water reflectance signal in the UV, does therefore not require any assumption on the water inherent optical properties. Based on the same observations, the 412 nm channel has previously been used with success in an atmospheric correction scheme applied over the Chesapeake region [39].

The ultraviolet spectral domain is also essential for assessing information about the colored dissolved organic matter distribution as already illustrated by recent works [111, 112]. The advantage of using nLw(325)/nLw(565) have been for instance emphasized by the latter authors for specifically assessing a_{CDOM} coefficient. Specifically, the interest of using of the UV reflectance signal stands in its ability for minimizing the potential issues represented by the overlapping of CDOM and phytoplankton absorption spectra occurring for the classically used blue and green bands. Indeed, in the UV domain CDOM is expected to dominate the absorption budget of the non-water compounds while the contribution of phytoplankton photoprotective compounds such as microsporine-like amino acids [113], although still uncertain, is expected to be relatively restricted in this spectral region. In addition to a gain in CDOM retrieval accuracy, the use of successive UV bands for deriving CDOM absorption from the marine reflectance signal would provide the opportunity to derive information on CDOM spectral slope which might provide innovative information on the nature of CDOM [114, 112].

Besides, reflectances in the near infrared domain have also been demonstrated to be of particular interest for estimating the SPM concentration from satellite imagery [e.g. AVHRR: [115]; MODIS, MERIS SeaWiFS: [116]]. In this spectral domain, direct relationships between NIR-reflectance and SPM values appear indeed to be robust due to the very low influence of CDOM absorption at these wavelengths. The latter reason has also led to the development of a variety of Chla inversion algorithms based on red and NIR wavelengths [117-122] allowing to avoid issues related to the use of classical blue/green reflectance ratios and taking in some cases advantages of the Chla fluorescence properties in the latter spectral domains.

2.　Directional polarized measurements

While the UV and NIR spectral domains are excellent candidates to overcome the challenges of ocean color remote sensing of coastal areas, the use of directional polarized radiometric measurements in the visible should also be examined. The inverse methods used to estimate the optical (IOPs) and biogeochemical parameters (i.e. Chl, POC, etc) information from space are based on different assumptions and mathematical developments. However, they all use the total remote sensing reflectance, $R_{rs}(\lambda)$, as input parameters (or a similar radiometric quantity). While theoretical and field measurements have highlighted that the polarization of the underwater light field is sensitive to the nature of the suspended marine particles (for example phytoplankton vs. mineral), none of these models exploits the linear polarization of the upwelling light field from the ocean surface. This is due to the fact that the marine polarized signal only represents a small fraction of the signal measured from space (about 10% of the total marine signal) which makes this measurement very challenging. Recent studies have however showed that space measurement of the polarized light field in different directions could provide useful information for both advanced atmospheric correction and bio-optical algorithms [123, 124]. For instance, the signal measured from the POLarization and Directionality of the Earth's Reflectances sensor (POLDER-2) over turbid areas has been found in excellent agreement with theory with regards to its variability with the bulk particulate matter [123]. However, the exploitation of such measurements in the frame of algorithms development for the futures ocean color missions requires an excellent radiometric resolution.

3.   Hyperspectral measurements

Hyperspectral sensors cover the visible and near infrared spectra with a much more complete spectral resolution, compared to multispectral ocean color sensors which have about 8 (i.e. MODIS) to 15 (i.e. MERIS) spectral bands in this spectral domain. For instance, the Hyperspectral Imager for the Coastal Ocean (HICO), operating on the international Space Station (ISS), specifically samples the coastal ocean every 5.7 nm [125]. Such high spectral resolution, associated with a high signal-to-noise ratio, allow the spectral features of the observed scene to be resolved. The interest of hyperspectral measurements compared to multispectral measurements for spatial ocean color applications is still under debate. However, more and more studies emphasize the importance of hyperspectral remote sensing data for phytoplankton detection, optically significant water constituents assessment, as well as bottom characterization [126-130]. Most of these studies are based on the derivative spectroscopy technique which enhances subtle spectral features present over hyperspectral data. While the second derivative of the radiometric signal allows specific spectral features to be detected, it is also much less sensitive to the presence of additional material such as colored dissolved organic matter which generally biases the standard algorithms based on a restricted number of wavelengths [128]. Futures ocean color sensors such as the HYPERSPECTRAL IMAGER (HSI) on board EnMAP will offer the opportunity to go further in the exploitation of hyperspectral data for ocean color applications.

4.   Advanced statistical methods

Due to the complexity of the coastal waters, the estimation of the water contents are more and more difficult to obtain with a high accuracy as shown in the previous sections. Therefore, there is a need in developing innovative inversion methods. Advanced statistic methods that can be very promising for the monitoring of the ocean color are the machine learning. These methods, developed initially for computational/artificial intelligence, are now more and more used in environmental sciences. As they have shown their utility in climate, meteorology, atmospheric sciences, satellite data processing, analysis and modeling of environmental data, weather prediction, they can be very helpful for studying the ocean color. Several studies already exist for the application of artificial neural networks for the atmospheric correction processing [50-52, 20, 57, 41], the estimation of the chlorophyll-a concentration [54, 131-136], the inherent optical properties retrieval [33, 137-138] or the diffuse attenuation coefficient [139]. But there exists other types of machine learning that the community could use such as the support vector machine [140, 141], the unsupervised competitive learning [142], regression trees [143], non-linear principal component [144], canonical correlation analysis [145]. These methods can be used for regression analysis for the direct estimation of water constituents, for classification (water types, phytoplankton functional groups) and for analysis of time series. For instance, Figure 15 shows a comparison of different algorithms for the estimation of the spectral diffuse attenuation coefficient $K_d(490)$ [139]. The use of a NN (bottom right Figure) allows to decrease the error for the estimation of moderate and high values of $K_d(490)$ ($K_d(490)>$ $0.5\ m^{-1}$).

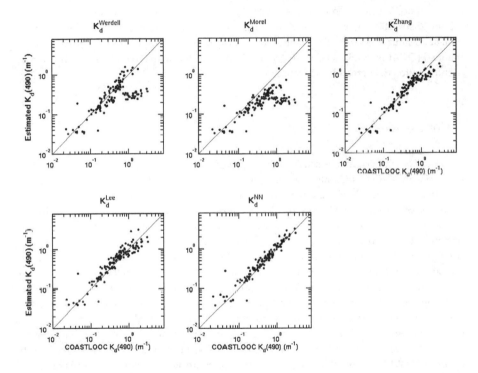

Figure 15. Scatterplots of desired $K_d(490)$ versus estimated $K_d(490)$ values retrieved with five algorithms using a coastal data set gathering measurements collected in coastal waters all around Europe [77]. The continuous line represents the 1:1 line. Figure from [139].

5.   High spatial and temporal resolutions satellite data

A high spatial resolution is required for ocean color remote sensing of coastal waters to catch the high spatial variability generally occurring in these areas, as well as for validation purposes. The recent development of geostationary ocean colour sensors will increase the precision of the remote sensing measurements and will provide relevant insights for the study of marine biogeochemical cycles [146, 147]. Geostationary satellites continuously view the same region of the earth's surface. It thus allows obtaining high quality and frequent observations of a defined area. Such an instrument is therefore particularly useful in order to follow the response of marine ecosystems to short-term variations in the environmental conditions. In particularly, it is of interest for monitoring the effects of rivers plumes and tidal front and mixing on the biotic and abiotic material present in coastal areas or assessing the effects of exceptional events (storms, red tides, dissemination of sediments or pollutants). It will also provide relevant information for biogeochemical modelling purposes as well as for research activities related to the biogeochemical cycles at daily scales. South Korea's instrument on board the COMS-1 satellite (GOCI, Geostationary Ocean Color Imager), which has been recently launched (2010), is the first ocean-colour sensor in a geostationary orbit. The target area of GOCI covers a large region (2500x2500 km) around the Korean peninsula. Its resolution is 500 m while it acquires data at a 1 hour frequency. The other ocean color geostationary missions which are currently planned (OCAPI-CNES, GeoCape-NASA) will increase the spatial coverage and the number of information delivered by such sensors.

5. Conclusions

The use of ocean color remote sensing in coastal waters offers a capability to routinely monitor the surface water constituents over large areas. Thanks to our better understanding of the remote sensing signal, as well as to the improvement sensor technology, new innovative products, compared to the historical chlorophyll concentration, are now available, at least at regional scales. One may cite for instance the colored dissolved organic matter, and concentration and nature of suspended particulate matter. Such information coupled with other data coming from other satellite observations (sea surface temperature, wind speed and direction, sea surface elevations) and physical and ecological modeling provide essential inputs to understand the complex bio-physical coupling occurring in coastal waters. For instance, the coupling between bottom sediment resuspension and observed surface suspended particulate matter has been stressed using satellite and physical modeling [148]. In the same way, based on the synergic use of satellite products deriving from ocean color observations, thermal measurements, and sea surface height, tight coupling bio-physical processes can now be examined [149]. While numerous ocean color products are now available in coastal areas, their assessment is not as accurate as for open ocean waters. Note that there is still no consensus of the community for a common algorithm to assess the chlorophyll concentration in coastal areas. One of the major issues is the retrieval of the marine radiometric signal from the top of atmosphere (i.e. the atmospheric corrections). Numerous new methods, listed in the present chapter, are now developed to address this particular point. A second issue is related to the bio-optical algorithms development. In the frame of algorithms development, new paths, compared to the ones used for open ocean waters, are still in the experimental phase. Among them, the simultaneous retrieval of atmospheric and oceanic components, the algorithms based on the classification approaches, as well as the exploitation of geostationary ocean color sensors should open new ways in a recent future.

Acknowledgements

This work has been supported by the ANR programme in the frame of the GlobCoast ANR-11-BLANC-BS56-018_01 project (www.foresea.fr/globcoast/)

Author details

Hubert Loisel[1,2,3], Vincent Vantrepotte[2], Cédric Jamet[2] and Dinh Ngoc Dat[3]

1 Laboratoire d'études en Géophysique et océanographie spatiales (LEGOS) CNRS: UMR5566 – IRD – CNES – Observatoire Midi-Pyrénées – INSU – Université Paul Sabatier - Toulouse, France

2 Laboratoire d'Océanologie et des Géosciences (LOG), INSU-CNRS, UMR 8187, Université Lille Nord de France, ULCO, Wimereux, France

3 Space Technology Institute (STI), Vietnam Academy of Science & Technology (VAST). Hanoi, Vietnam

References

[1] Taylor AH, Allen JI, Clark PA. Extraction of a weak climatic signal by an ecosystem. Nature 2002; 416 629-632.

[2] Hays CG. Effect of Nutrient Availability, Grazer Assemblage and Seagrass Source Population on the Interaction Between Thalassia Testudinum (Turtle Grass) and its Algal Epiphytes. Journal of Experimental Marine Biology and Ecology 2005; 314 53-68.

[3] Paerl, HW. Assessing and Managing Nutrient-Enhanced Eutrophication in Estuarine and Coastal Waters: Interactive Effects of Human and Climatic Perturbations. Ecological Engineering 2006; 26 40-54.

[4] Wollast R. Evaluation and Comparison of the Global Carbon Cycle in the Coastal Zone and in the Open Ocean. In: K. H. Brink and A. R. Robinson (eds.). The Global Coastal Ocean. John Wiley & Sons; 1998. p213-252.

[5] Naqvi, S.W.A.; Unnikrishnan, A.S. 2009. Hydrography and biogeochemistry of the coastal ocean In: Le Quere C, Saltzman E. (eds.) Surface ocean - Lower atmosphere processes. AGU; 2009. p233-250.

[6] auer, JE, Druffel ERM. Ocean Margins as a Significant Source of Organic Matter to the Deep Open Ocean. Nature 1998; 392 482-485.

[7]   Hedges JL. Global Biogeochemical Cycles: Progress and Problems. Marine Chemistry 1992; 39 67-93.

[8]   Gordon HR, Wang M. Retrieval of Water-Leaving Radiance and Aerosol Optical Thickness Over the Oceans with SeaWiFS: A Preliminary Algorithm. Applied Optics 1994; 33 443-452.

[9]   Antoine D, Morel A. A Multiple Scattering Algorithm for Atmospheric Correction of Remotely-Sensed Ocean Colour (MERIS Instrument): Principle and Implementation for Atmospheres Carrying Various Aerosols Including Absorbing Ones. International Journal of Remote Sensing 1999; 20(9) 1875-1916.

[10] Morel A., editor. Minimum Requirements for an Operational Ocean-Colour Sensor for the Open Ocean. International Ocean Color Coordinating Group; 1998.

[11] Gordon HR, McCluney WR. Estimation of the Depth of Sunlight Penetration in the Sea for Remote Sensing. Applied Optics 1975, 14 413-416.

[12] Smith RC, Baker KS. Optical Classification of Natural Waters; Limnology & Oceanography 1978; 23(2) 260-267.

[13] McClain CR. A Decade of Satellite Ocean Color Observations. Annual Review of Marine Science 2009; 1 19–42.

[14] Stramski D, Reynolds RA, Babin M, Kaczmarek S, Lewis MR, Röttgers R, Sciandra A, Stramska M, Twardowski MS, Claustre H. Relationships Between the Surface Concentration of Particulate Organic Carbon and Optical Properties in the Eastern South Pacific and Eastern Atlantic Oceans. Biogeosciences 2008; 4 1041-1058.

[15] Loisel H, Nicolas JM, Deschamps PY, Frouin R. Seasonal and Inter-Annual Variability of the Particulate Matter in the Global Ocean. Geophysical Research Letters 2002; 29(24) DOI:10.1029/2002GLO15948.

[16] Duforet-Gaurier L, Loisel L, Dessailly D, Nordkvist K, Alvain, S. Estimates of Particulate Organic Carbon over the Euphotic Depth from In Situ Measurements. Application to Satellite Data over the Global Ocean. Deep-Sea Research I 2010; 57 351–367

[17] Lee Z-P., editor. Remote Sensing of Inherent Optical Properties: Fundamentals, Tests of Algorithms and Applications. International Ocean Color Coordinating Group; 2006.

[18] Loisel, Nicolas J-M, Sciandra A, Stramski D, Poteau A. Spectral Dependency of Optical Backscattering by Marine Particles from Satellite Remote Sensing of the Global Ocean. Journal of Geophysical Research 2006; 111(C9) C0902410.1029/2005JC003367.

[19] Bricaud A, Ciotti AM, Gentili B. Spatial-Temporal Variations in Phytoplankton Size and Colored Detrital Matter Absorption at Global and Regional Scales, as Derived from Twelve Years of SeaWiFS Data (1998-2009). Global Biogeochemical Cycles 2012; 26 doi:10.1029/2010GB003952.

[20] Jamet C, Thiria S, Moulin, Crepon, M. Use of a Neuro-Variational Inversion for retrieving Oceanic and Atmospheric Constituents from Ocean Color Imagery: a Feasibility Study. Journal of Atmospheric and Oceanic Technology 2005; 22(4) 460-475.

[21] Sathyendranath S, Watts L, Devred E, Platt T, Caverhill C, Maass H. Discrimination
of Diatoms from Other Phytoplankton using Ocean-Colour Data. Marine Ecology
Progress Series 2005; 272 59–68.

[22] Alvain S, Moulin C, Dandonneau Y, Breon F-M. Remote Sensing of Phytoplankton
Groups in Case 1 Waters from Global SeaWiFS imagery. Deep Sea Research Part I
2005; 52(11) 1989–2004.

[23] Ciotti A, Bricaud A. Retrievals of a Size Parameter for Phytoplankton and Spectral
Light Absorption by Colored Detrital Matter from Water-Leaving Radiances at Sea-
WiFS Channels in a Continental Shelf Region off Brazil, Limnology 1 Oceanography
Methods 2006; 4 237–253.

[24] Kostadinov TS, Siegel DA, Maritorena S. Retrieval of the Particle Size Distribution
from Satellite Ocean Color Observations. Journal of Geophysical Research 2009; 114
doi:10.1029/2009JC005303.

[25] International Ocean Color Coordinating Group. http:www.ioccg.org (accessed 14 October 2012).

[26] Alvain S, Moulin C, Dandonneau H, Loisel H. Seasonal Distribution and Succession
of Dominant Phytoplankton Groups in the Global Ocean: A satellite view. Global Biogeochemical Cycles 2008; 22(3) DOI: 10.1029/2007GB003154.

[27] Sathendranath S., editor. Remote sensing of Ocean Colour in coastal and Other Opti-
cally-Complex, Waters. International Ocean Color Coordinating Group; 2003.

[28] Gordon HR. Atmospheric Correction of Ocean Color Imagery in the Earth Observing
System Era. Journal of Geophysical Research 1997;102(D14) 17081-17106.

[29] Robinson, IS. Satellite Observations of Ocean Colour, Philosophical Transactions of
the Royal Society of London Series A- Mathematical Physical and Engineering Sciences 1983;309(1508) 338-347.

[30] Siegel DA, Wang MH, Maritorena S, Robinson W. Atmospheric Correction of Satellite Ocean Color Imagery: the Black Pixel Assumption. Applied Optics 2000;39(21)
3582-3591.

[31] Bailey SW, Franz BA, Werdell PJ. Estimation of Near-Infrared Water-Leaving Reflectance for Satellite Ocean Color Data Processing. Optics Express 2010;18 7521–7527.

[32] Brajard, J, Jamet C, Moulin C, Thiria S. Use of a Neuro-Variational Inversion for Retrieving Oceanic and Atmospheric Constituents from Satellite Ocean Colour Sensor:
Application to Absorbing Aerosols. Neural Networks 2006;19 178–185.

[33] Doerffer R, Schiller H. The MERIS Case 2 Water Algorithm. International Journal of Remote Sensing 2007;28 517–535.

[34] Dogliotti A-I, Ruddick K, Nechad B, Lasta C, Mercado A, Hozbor C, Guerrero R, Riviello López G, Abelando M. Calibration and validation of an algorithm for remote sensing of turbidity over La Plata River estuary, Argentina. In: Reuter R (eds) EARSeL eProceedings, 5th EARSeL Workshop on Remote Sensing of the Coastal Zone, 1st - 3rd June 2011, Prague, Czech Republic. Oldenbrug: BIS-Verlag; 2011.

[35] Hu C, Carder KL, Muller-Karger FE. Atmospheric Correction of SeaWiFS Imagery over Turbid Coastal Waters. Remote Sensing of Environment 2000; 74 195–206.

[36] Kuchinke CP, Gordon HR, Franz BA. Spectral Optimization for Constituent Retrieval in Case II Waters I: Implementation and Performance. Remote Sensing of Environment 2009;13 571–587.

[37] Land PE, Haigh, JD. Atmospheric Correction over Case 2 Waters with an Iterative Fitting Algorithm. Applied Optics 1996;35 5443–5451.

[38] Lavender SJ, Pinkerton MH, Moore GF, Aiken J, Blondeau-Patissier D. Modification to the Atmospheric Correction of SeaWiFS Ocean Colour Images over Turbid Waters. Continental Shelf Research 2005; 25 539–555.

[39] Oo, M, Vargas M, Gilerson A, Gross B, Moshary F, Ahmed S. Improving Atmospheric Correction for Highly Productive Coastal Waters using the Short Wave Infrared Retrieval Algorithm with Water-Leaving Reflectance Constraints at 412 nm. Applied Optics 2008;47(21) 3846–3859.

[40] Ruddick K, Ovidio F, Rijkeboer M. Atmospheric Correction of SeaWiFS Imagery for Turbid Coastal and Inland Waters. Applied Optics 2000;39(6) 897–912.

[41] Schroeder T, Behnert I, Schaale M, Fischer J, Doerffer R. Atmospheric Correction Algorithm for MERIS above Case-2 Waters. International Journal of Remote Sensing 2007;28 1469–1486.

[42] Shanmugam, P, Ahn, YH.. New Atmospheric Correction Technique to Retrieve the Ocean Colour from SeaWiFS Imagery in Complex Coastal Waters. Journal of Optics A: Pure and Applied Optics 2007;9 511–530.

[43] Stumpf RP, Arnone RA, Gould RW, Ransibrahmanakul V. A Partly Coupled Ocean–Atmosphere Model for Retrieval of Water-Leaving Radiance from SeaWiFS in Coastal Waters. NASA technical memorandum 2003; 206892(22).

[44] Wang MH, Shi S. Estimation of Ocean Contribution at the MODIS Near-Infrared Wavelengths along the East Coast of the US: Two Case Studies. Geophysical Research Letters 2005;32(13) DOI: 10.1029/2005GL022917

[45] Wang, M, Son S, and Shi W. Evaluation of MODIS SWIR and NIR-SWIR Atmospheric Correction Algorithms using SeaBASS Data, Remote Sensing of Environment 2009; 113 635-644.

[46] Wang MH, Shi W, Jiang LD. Atmospheric Correction using Near-Infrared Bands for Satellite Ocean Color Data Processing in the Turbid Western Pacific region. Optics Express 2012; 20(2) 741-753.

[47] Wang MH, Shi W. The NIR-SWIR Combined Atmospheric Correction Approach for MODIS Ocean Color Data Processing. Optics Express 2007;15 15722-15733.

[48] Shi W, Wang MH. An Assessment of the Black Ocean Pixel Assumption for MODIS SWIR Band. Remote Sensing of Environment 2009;113 1587-1597.

[49] Werdell PJ, Franz BA, Bailey SW. Evaluation of Shortwave Infrared Atmospheric Correction for Ocean Color Remote Sensing of Chesapeake Bay. Remote Sensing of Environement 2010; 114 2238-2247.

[50] Brajard J, Jamet C, Moulin C, Thiria S. Validation of a Neuro-Variational Inversion of Ocean Colour Images. Advances in Space Research 2006;38 2169–2175.

[51] Brajard, J, Moulin C, Thiria S. Atmospheric Correction of SeaWiFS Ocean Color Imagery in the Presence of Absorbing Aerosols off the Indian Coast using a Neuro-Variational Method. Geophysical Research Letters 2008;35 doi:10.1029/2008GL035179.

[52] Brajard J, Santer R, Crépon M, Thiria S. Atmospheric Correction of MERIS Data for Case-2 Waters using a Neuro-Variational Inversion. Remote Sensing of Environment 2012;126 51-61.

[53] Chomko, RM, Gordon HR. Atmospheric Correction of Ocean Color Imagery: Use of the Junge Power-Law Aerosol Size Distribution with Variable Refractive Index to Handle Aerosol Absorption. Applied Optics 1998;37 5560-5572.

[54] Schiller H, Doerffer R. Neural Network for the Evaluation of the Inverse Model: Operational Derivation of Case 2 Water Properties from MERIS Data. International Journal of Remote Sensing 1997;20 1735–1746.

[55] Bishop CM., editor. Neural Networks for Pattern Recognition. Oxford University Press; 1995.

[56] Jamet, C., and H. Loisel. 2009. Data Assimilation in Surface Ocean/Lower Atmosphere Processes. Geophysical Monograph 187. Editors C. Le Quere and E. Saltzman. AGU (pp 329).

[57] Jamet, C, Moulin, Thiria S. Monitoring Aerosol Optical Properties over the Mediterranean from SeaWiFS Images using a Neural Network Inversion. Geophysical Research Letters 2004; 31 doi:10/1029/2004GL019951.

[58] Jamet C, Loisel H, Kuchinke CP, Ruddick K, Zibordi G, Feng H. Comparison of Three SeaWiFS Atmospheric Correction Algorithms for Turbid Waters using AERO-NET-OC Measurements. Remote Sensing of Environment 2011;115(8) 1955-1965.

[59] Goyens C, Jamet C, Ruddick K. Validation of Marine Models for Visible and Near In-fra-Red Wavelengths as used for Turbid Water Atmospheric Correction. Optics Express 2012; submitted.

[60] Loisel H, Stramski D. Estimation of the Inherent Optical Properties of Natural Waters from Irradiance Attenuation Coefficient and reflectance in the Presence of Raman Scattering. Applied Optics 2000; 39(18): 3001-3011.

[61] Goodman JA, Lee Z-P, Ustin SL. Influence of Atmospheric and Sea-Surface Correc-tions on Retrieval of Bottom Depth and Reflectance using a Semi-Analytical Model: a Case Study in Kaneohe Bay, Hawaii. Applied Optics 2008;47 F1-F11.

[62] Mobley CD. Zhang H, Voss K. Effects of Optically Shallow Bottoms on Upwelling Radiances: Bidirectional Reflectance Distribution Function Effects. Limnology and Oceanography 2003;48 337-345.

[63] Ohde T, Siegel H. Correction of Bottom Influence in Ocean Colour Satellite Images of Shallow Water Areas of the Baltic Sea. International Journal of Remote Sensing 2001;22 297-313.

[64] Tanré D, Herman M, Deschamps P-Y. Influence of the Background Contribution upon Space Measurements of Ground Reflectance. Applied Optics 1981;20 3676–3684.

[65] Tanré D, Herman M, Deschamps P-Y, de Leffe A. Atmospheric Modeling for Space Measurements of Ground Reflectances, including Bidirectional Properties. Applied Optics 1979;18 3587–3594.

[66] Santer R, Schmechtig C. Adjacency Effects on Water Surfaces: Primary Scattering Ap-proximation and Sensitivity Study. Applied Optics 2000; 39(3) 361-375.

[67] Ahmad Z, Franz BA, McClain CR, Kwiatkowska EJ, Werdell J, Shettle EP, Holben BN. New Aerosol Models for the Retrieval of Aerosol Optical Thickness and Normal-ized Water-Leaving Radiances from the SeaWiFS and MODIS Sensors over Coastal Regions and Open Oceans. Applied Optics 2010; 49(29) 5545–5560.

[68] Moulin C, Gordon HR, Banzon VF, Evans RH. Assessment of Saharan Dust Absorp-tion in the Visible from SeaWiFS Imagery. Journal of Geophysical Research 2001;106 18239-18249.

[69] Banzon VF, Gordon HR, Kuchinke CP, Antoine D, Voss KJ, Evans RH. Validation of a SeaWiFS Dust-Correction Methodology in the Mediterranean Sea: Identification of an Algorithm-Switching Criterion. Remote Sensing of Environment 2009; 113 2689-2700.

[70] Platt T, Hoepffner N, Stuart V, Brown C., editors. Why Ocean Colour? The Societal Benefits of Ocean- Colour Technology. International Ocean Color Coordinating Group; 2008.

[71] Hu C, Muller-Karger FE, Taylor C, Carder KL, Kelble C, Johns E, Heil CA. Red Tide Detection and Tracing using MODIS Fluorescence Data: A Regional Example in SW Florida Coastal Waters. Remote Sensing of Environment 2005;97 311 – 321.

[72] Ahn Y-H, Shanmugam P. Detecting the Red Tide Algal Blooms from Satellite Ocean Color Observations in Optically Complex Northeast-Asia Coastal Waters. Remote Sensing of Environment 2006; 103 419-437.

[73] Ishizaka J, Kitaura Y, Touke Y, Sasaki H, Tanaka A, Murakami H, Suzuki T, Matsuoka K, Nakata H. Satellite Detection of Red Tide in Ariake Sound, 1998-2001. Journal of Oceanography 2006;62 37-45.

[74] Mannino A, Russ ME, Hooker SB. Algorithm Development and Validation for Satellite-Derived Distributions of DOC and CDOM in the U.S. Middle Atlantic Bight. Journal of Geophysical Research 2008;113 doi:10.1029/2007JC004493.

[75] Del Castillo CE, Miller RL. On the Use of Ocean Color Remote Sensing to Measure the Transport of Dissolved Organic Carbon by the Mississippi River plume. Remote Sensing of Environment 2008;112 836–844.

[76] Matsuoka A, Bricaud A, Benner R, Para J, Sempéré R, Prieur L, Bélanger S, Babin M. Tracking the Transport of Colored Dissolved Organic Matter in the Southern Beaufort Sea: Relationship with Hydrographic Characteristics. Biogeosciences 2012;9 925–940.

[77] Babin M, Stramski D, Ferrari GM, Claustre H, Bricaud A, Obolensky G, Hoepffner N. Variations in the Light Absorption Coefficients of Phytoplankton, Nonalgal Particles, and Dissolved Organic Matter in Coastal Waters around Europe. Journal of Geophysical Research 2003;108(C7) doi:10.1029/2001JC000882.

[78] Loisel H, Mériaux X, Berthon J-F, Poteau A. Investigation of the Optical Backscattering to Scattering Ratio of Marine Particles in Relation to their Biogeochemical Composition in the Eastern English Channel and southern North Sea. Limnology and Oceanography 2007;52(2) 739-752.

[79] Vantrepotte V, Brunet C, Mériaux X, Lécuyer E, Vellucci V, Santer R. Bio-Optical Properties of Coastal Waters in the Eastern English Channel. Estuarine Coastal and Shelf Science 2007; 72(1-2) 201-212.

[80] Darecki M, Stramski D. An Evaluation of MODIS and SeaWiFS Bio-Optical Algorithms in the Baltic Sea. Remote Sensing of Environment 2004;89 326-350.

[81] Loisel H, Lubac B, Dessailly D, Duforêt-Gaurier L, Vantrepotte V. Effect of Inherent Optical Properties Variability on the Chlorophyll Retrieval from Ocean Color Remote Sensing: an In Situ Approach. Optics Express 2010; 18(20) 2010.

[82] Bailey S, Wang M. Satellite aerosol optical thickness match-up procedures. NASA technical memorandum 2001; 2001-209982 70–72.

[83] Feng H, Campbell JW, Dowell MD, Moore TS. Modeling Spectral Reflectance of Optically Complex Waters using Bio-Optical Measurements from Tokyo Bay. Remote Sensing of Environment 2005;99(3) 232–243.

[84] Vantrepotte V, Loisel H, Mériaux X, Neukermans G, Dessailly D, Jamet C, Gensac E, Gardel A. Seasonal and Inter-Annual (2002-2010) Variability of the Suspended Particulate Matter as Retrieved from Satellite Ocean Color Sensor over the French Guiana Coastal Waters. Journal of Coastal Research 2011;64 1750-1754.

[85] Robinson WD, Franz BA, Patt FS, Bailey SW, Werdell PJ. Masks and Flags Updates. NASA Technical Memorandum 2003;22 34-40.

[86] Wang M., Shi W. Cloud Masking for Ocean Color Data Processing in the Coastal Regions. IEEE Transaction on Geoscience & Remote Sensing 2006;44 3196-3205.

[87] Nicolas J-M, Deschamps P-Y, Loisel H, Moulin C. Algorithm Theoretical Basis Document, POLDER-2 / Ocean Color / Atmospheric corrections. http://smsc.cnes.fr/POLDER/A_produits_scie.htm (accessed 17 October 2012).

[88] Nordkvist K, Loisel H, Duforêt-Gaurier L. Cloud Masking of SeaWiFS Images over Coastal Waters using Spectral Variability. Optics Express 2009; 17(15) 12246 12258.

[89] Feldman GC, McClain CR. Ocean Color Web, SeaWiFS Reprocessing 5.2. http://oceancolor.gsfc.nasa.gov.

[90] He X, Bai Y, Pan D, Tang J, Wang D. Atmospheric Correction of Satellite Ocean Color Imagery using the Ultraviolet Wavelength for Highly Turbid Waters. Optics Express 2012;20(18) 20754-20770.

[91] Ruddick K, De Cauwer V, Park Y, Moore G. Seaborne Measurements of Near Infrared Water-Leaving Reflectance: The Similarity Spectrum for Turbid Waters. Limnology and Oceanography 2006; 51(2) 1167–1179.

[92] Doron M, Bélanger S, Doxaran D, Babin M. Spectral Variations in the Near-Infrared Ocean Reflectance. Remote Sensing of Environment 2011; 115(7) 1617-1631.

[93] D'Sa EJ, Miller RL, Del Castillo C.. Bio-optical Properties and Ocean Color Algorithms for Coastal Waters Influenced by the Mississippi River during a Cold Front. Applied Optics 2006;45 7410-7428.

[94] Garcia VMT, Signorini SR, Carcia CAE, McClain CR. Empirical and Semi-Analytical Chlorophyll Algorithms in the South-Western Atlantic Coastal Region (25−40°S and 60−45°W). International. Journal of Remote Sensing 2006; 27(8) 1539-1562.

[95] Kowalczuk P, Darecki M, Zabłocka M, Górecka I. Validation of Empirical and Semi-Analytical Remote Sensing Algorithms for Estimating Absorption by Colored Dis-

solved Organic Matter in the Baltic Sea from SeaWiFS and MODIS Imagery. Oceanologia 2010;, 55(2) 171-196.

[96] Baker KS, Smith RC. Bio-Optical Classification and Model of Natural Waters. 2. Limnology and Oceanography 1982; 27(3) 500-509.

[97] Prieur L., Sathyendranath S. An Optical Classification of Coastal and Oceanic Waters based on the Specific Spectral Absorption Curves of Phytoplankton Pigments, Dissolved Organic Matter, and other Particulate Materials. Limnology and Oceanography 1981;26 671–689.

[98] Reinart A, Herlevic A, Arstb H, Sipelgas L. Preliminary Optical Classification of Lakes and Coastal Waters in Estonia and South Finland. Journal of Sea Research 2003;49 357-366.

[99] Moore TS, Campbell JW, Feng H. A fuzzy logic classification scheme for selecting and blending satellite ocean color algorithms. IEEE Transactions on Geoscience and Remote Sensing 2001; 39(8) 1764-1776.

[100] Moore TS, Campbell JW, Dowell MD. A Class-Based Approach to Characterizing and Mapping the Uncertainty of the MODIS Ocean Chlorophyll Product. Remote Sensing of Environment 2009;113 2424-2430.

[101] D'Alimonte D, Mélin F, Zibordi G, Berthon, J-F. Use of the Novelty Detection Technique to Identify the Range of Applicability of the Empirical Ocean Color Algorithms. IEEE Transactions on Geoscience and Remote Sensing 2003;41 2833–2843.

[102] Szeto M, Moore TS, Campbell JW, Werdell PJ. Are the World's Oceans Optically Different, Journal of Geophysical Research-Oceans 2011;116. doi: 10.1029/2011JC007230.

[103] Forget M-H, Stuart V, Platt T., editors. Remote Sensing in Fisheries and Aquaculture. International Ocean Color Coorditaning Group; 2009.

[104] Lubac B, Loisel H.. Variability and Classification of Remote Sensing Reflectance Spectra in the Eastern English Channel and Southern North Sea. Remote Sensing of Environment 2007;110, 45–58.

[105] Tilstone GH, Angel-Benavides IM, Pradhan Y, Shutler JD, Groom S, Sathyendranath S. An Assessment of Chlorophyll-a Algorithms Available for SeaWiFS in Coastal and Open Areas of the Bay of Bengal and Arabian Sea. Remote Sensing of Environment 2011;115 2277–2291

[106] Hommersom A, Wernand MR, Peters S, Eleveld MA, van der Woerd HJ, de Boer J.. Spectra of a Shallow Sea - Unmixing for Class Identification and Monitoring of Coastal Waters. Ocean Dynamics 2011;61 463-480.

[107] Le C, LI Y, Zha Y, Sun D, Huang C, Zhang H. Remote Estimation of Chlorophyll a in Optically Complex Waters based on Optical Classification. Remote Sensing of Environment 2010;115(2) 725-737.

[108] Mélin F, Vantrepotte V, Clerici M, D'Alimonte D, Zibordi G, Berthon J.F. Multi-Sensor Satellite Time Series of Optical Properties and Chlorophyll a Concentration in the Adriatic Sea. Progress in Oceanography 2011;31 229-244.

[109] Vantrepotte V, Loisel H, Dessailly D, Mériaux X. Optical Classification of Contrasted Coastal Waters. Remote Sensing of Environment 2012;123 306–323.

[110] Hoge FE, Lyon PE, Swift RN, Yungel JK, Abbott MR, Letelier RM, Esaias WE. Validation of Terra-MODIS Phytoplankton Chlorophyll Fluorescence Line Height. I. Initial Airborne Lidar Results. Applied Optics 2003. 42(15) 2767-2771.

[111] Hooker SB, CR McClain, Mannino A. NASA Strategic Planning Document: A Comprehensive Plan for the Long-Term Calibration and Validation of Oceanic Biogeochemical Satellite Data. NASA Special Publication 2007; SP-2007-214152 1– 31

[112] Tedetti M, Charrière B, Bricaud A, Para J, Raimbault P, Sempéré R. Distribution of Normalized Water-Leaving Radiances at UV and Visible Wavebands in Relation with Chlorophyll a and Colored Detrital Matter Content in the South East Pacific. Journal of Geophysical Research 2010;115 doi:10.1029/2009JC005289.

[113] Vernet M, Whitehead K. Release of Ultraviolet-Absorbing Compounds by the Red-Tide dinoflagellate Lingulodinium Polyedra. Marine Biology 1996;127 35–44.

[114] Fichot, C. G., and R. Benner. A Novel Method to Estimate DOC Concentrations from CDOM Absorption Coefficients in Coastal Waters. Geophysical Research Letters 2011;38 doi:10.1029/2010GL046152.

[115] Myint, S.W., and N. Walker. Quantification of Surface Suspended Sediments Along a River Dominated Coast with NOAA AVHRR and SeaWiFS Measurements: Louisiana, USA. International Journal of Remote Sensing 2002. 23(16): 3229-3249.

[116] Nechad, B., Ruddick K.G. and Park Y. Calibration and validation of a generic multi-sensor algorithm for mapping of total suspended matter in turbid waters. Remote Sensing of Environment 2010, 114, 854-866.

[117] Gower, J., King, S. Validation of chlorophyll fluorescence derived from MERIS on the west coast of Canada. International Journal of Remote Sensing 2007, 28, 625–635

[118] Gons, H. -J., Rijkeboer, M., Ruddick, K. -G.. A chlorophyll-retrieval model for satellite imagery (Medium Resolution Imaging Spectrometer) of inland and coastal waters. Journal of Plankton Research, 2002, 24, 947–951.

[119] Gitelson, A. -A., Dall'Olmo, G., Moses, W., Rundquist, D., Barrow, T., Fisher, T. -R. A simple semi-analytical model for remote estimation of chlorophyll-a in turbid productive waters: Validation. Remote Sensing of Environment 2008, 112, 3582–3593.

[120] Gilerson, A., Gitelson, A. -A., Zhou, J., Gurlin, D., Moses, W. -J., Ioannou, I. Algorithms for remote estimation of chlorophyll-a in coastal and inland waters using red and near infrared bands. Optics Express 2011, 18, 24109–24125.

[121] Yang, W., Matsushita, B., Chen, J., Fukushima, T., Ma, R. An enhanced threeband index for estimating chlorophyll-a in turbid case-II waters: Case studies of Lake Kasumigaura, Japan, and Lake Dianchi, China. IEEE Geoscience and Remote Sensing Letters 2010, 7, 655–659.

[122] Gurlin, D., Gitelson, A. A., Moses, W. J. Remote estimation of chl-a concentration in turbid productive waters — Return to a simple two-band NIR-red model? Remote Sensing of Environment 2011; 115(12), 3479–3490.

[123] Loisel H, Duforet L, Dessailly D, Chami M, Dubuisson P. Investigation of the Variations in the Water Leaving Polarized Reflectance from the POLDER Satellite Data over two Biogeochemical Contrasted Oceanic Areas. Optics Express 2008; 16(17) 12905-12918.

[124] Harmel T, Chami M. Influence of Polarimetric Satellite Data Measured in the Visible Region on Aerosol Detection and on the Performance of Atmospheric Correction Procedure over Open Ocean Waters. Optics Express 2008; 19 (21), 20960-20983

[125] Lucke, R. L., M. Corson, N. R. McGlothlin, S. D. Butcher, D. L. Wood, D. R. Korwan, R. R. Li, W. A. Snyder, C. O. Davis, and D. T. Chen. Hyperspectral Imager for the Coastal Ocean: instrument description and first images. Applied Optics 2011; 50(10), 1501-1516.

[126] Lee, Z., K.L. Carder, C.D. Mobley, R.G. Steward, and J.S. Patch. Hyperspectral Remote Sensing for Shallow Waters. 2. Deriving Bottom Depths and Water Properties by Optimization. Applied Optics 1999; Vol. 38, Issue 18, pp. 3831-3843.

[127] Craig, S.E., S.E. Lohrenz, Z. Lee, K.L. Mahoney, G.J. Kirkpatrick, O.M. Schofield, and R.G. Steward. Use of hyperspectral remote sensing reflectance for detection and assessment of the harmful alga, Karenia brevis, Applied Optics. 2006; 45, 5414– 5425.

[128] Lubac, B., H. Loisel, N. Guiselin, R. Astoreca, L.F. Artigas, and X. Mériaux. Hyperspectral and multispectral ocean color inversions to detect Phaeocystis globosa blooms in coastal waters. Journal of Geophysical Research 2008; 113, C06026. doi: 10.1029/2007JC004451.

[129] Louchard, E.M., R.P. Reid, C.F. Stephens, C.O. Davis, R.A. Leathers, R. A., Downes, T. V., et al. Derivative analysis of absorption features in hyperspectral remote sensing data of carbonate sediments. Optics Express 2002; 10, 1573–1584.

[130] Torrecilla, E., D. Stramski, R. A. Reynolds, E. Millán-Núñez, and J. Piera. Cluster analysis of hyperspectral optical data for discriminating phytoplankton pigment assemblages in the open ocean. Remote Sensing of Environment 2011; 115, 2578-2593.

[131] Buckton, D., E. O'Mongain, and S. Danaher. The use of neural networks for the estimation of oceanic constituents based on the MERIS instrument. International Journal of Remote Sensing 1999; 20, 1841–1851.

[132] D'Alimonte, D., and G. Zibord. Phytoplankton determination in an optically complex coastal region using a multilayer perceptron neural network. IEEE Trans. Geosci. Remote Sensing 2003; 41, 2861–2868.

[133] Gonzales Vilas, L., E. Spyrakos, and J. M. Torres-Palenzuela. Neural network estimation of chlorophyll a from MERIS full resolution data for the coastal waters of Galician rias (NW Spain), Remote Sensing of Environment 2011; 115, 524–535.

[134] Gross, L., S. Thiria, R. Frouin, and B. G. Mitchell. Artificial neural network for modeling the transfer function between marine reflectance and phytoplankton pigment concentration. Journal of Geophysical Research 2000; 105, 3483–3495.

[135] Tanaka, A., M. Kishino, R. Doerffer, H. Schiller, T. Oishi, and T. Kubota. Development of a neural network algorithm for retrieving concentrations of chlorophyll, suspended matter and yellow substance from radiance data of the ocean color and temperature scanner, J. Oceanography 2004; 60, 519–530.

[136] Zhang, T., F. Fell, Z. S. Liu, R. Preusker, J. Fischer, and M. X. He. Evaluating the performance of artificial neural network techniques for pigment retrieval from ocean color in case I waters. Journal of Geophysical Research 2003; 108(C9), 3286, doi: 10.1029/2002JC001638.

[137] Bricaud, A., C. Mejia, D. Biondeau-Patissier, H. Claustre, M. Crepon, and S. Thiria. Retrieval of pigment concentrations and size structure of algal populations from their absorption spectra using multilayered perceptrons. Applied Optics 2007; 46, 1251–1260.

[138] Ioannou, I., A. Gilerson, B. Gross, F. Moshary, and S. Ahmed. Neural network approach to retrieve the inherent optical properties of the ocean from observations of MODIS. Applied Optics 2011; 50, 3168–3186.

[139] Jamet, C., H., Loisel and D., Dessailly. Retrieval of the spectral diffuse attenuation coefficient $Kd(\lambda)$ in open and coastal ocean waters using a neural network inversion. Journal of Geophysical Research 2012; 117, doi:10.1029/2012JC008076.

[140] Vapnik, V.N. (1995). The Nature of Statistical Learning Theory, Springer-Verlag Eds, Berlin.

[141] Vapnik, V.N. (1998). Statistical Learning Theory, Wiley Edzs, New York.

[142] Kohonen T. Self-Organizing Maps, Third extended edition. Springer; 2001.

[143] Breiman L, Friedman JH, Olshen RA, Stone CJ. Classification and Regression Trees. Wadsworth; 1983.

[144] Hsieh WW. Nonlinear Principal Component Analysis by Neural Networks. Tellus Series A-Dynamic Meteorology and Oceanography 2001;53(5) 599-615.

[145] Hsieh WW. Nonlinear Canonical Correlation Analysis by Neural Network. Neural Networks 2000;13(10) 1095-1105.

[146] Choi, J.-K., Y. J. Park, J. H. Ahn, H.-S. Lim, J. Eom, and J.-H. Ryu. GOCI, the world's first geostationary ocean color observation satellite, for the monitoring of temporal variability in coastal water turbidity. Journal of Geophysical Research 2012; 117, C09004, doi:10.1029/2012JC008046.

[147] Lei, M., Roman, A., Bricaud, A., Froidefond, J.M., Mathieu, S., and P. Gouton. Simulation of future geostationary ocean color images. IEEE Journal of Selected Topics in Applied Earth Observations and Remote Sensing 2012; 5, 1, 173-182.

[148] Ouillon S., P. Douillet, and S. Andréfouët. Coupling satellite data with in situ measurements and numerical modeling to study fine suspended sediment transport: a study for the lagoon of New Caledonia, Coral Reefs 2004; 23 (1), 109-122.

[149] Arnon, R.A., and A.R. Parsons. Real-time use of ocean color remote sensing for coastal monitoring. Remote Sensing of Coastal Aquatic Environments 2005; Editors R.L. Miller et al. Springer (pp 345).

Chapter 4

Near Surface Turbulence and Gas Exchange Across the Air-Sea Interface

Qian Liao and Binbin Wang

Additional information is available at the end of the chapter

1. Introduction

Oceans and the atmosphere are tightly coupled to influence the energy budget, gas exchange and the global climate. For example, 20%~30% anthropogenic emitted CO_2 was sequestered by oceans. The ocean surface boundary layer plays an intermediate role on the exchange of mass, momentum and energy between air and ocean. Quantifying transport terms (such as temperature, gas fluxes) across the air-water interface has been an important subject of physical oceanography for several decades.

The physical mechanism of interfacial gas exchange is essentially natural and complex for several reasons. (1) Fluid motions on both sides of the interface are typically turbulent, which demonstrate a stochastic feature with a wide range of scales; (2) The interface can be disturbed and hence deformed into irregular shapes, interacts with the turbulence structure in both boundary layers.

Gas transfer velocity k is the key parameter that regulates the interfacial gas exchange, which is usually determined through experimental methods. The gas transfer velocity can be defined as

$$k = \frac{F}{C_w - \alpha C_a} \qquad (1)$$

where, F is gas flux across the air water interface, C_w and C_a are bulk gas concentration at water and air phase, α is the Ostwald solubility coefficient.

Although the definition is simple, quantification of gas transfer velocity is indirect and difficult since it is affected by a wide range of environmental variables, including wind, rainfall, surfactants, surface waves, etc. Most of these variables are strongly coupled to affect the gas transfer process.

2. Theoretical models

It is well accepted that the gas flux of slightly soluble (such as CO_2) across the air-water interface without wave breaking is largely controlled by the transport mechanism through a very thin aqueous diffusive boundary layer. The gas transfer velocity is determined by molecular transport though this thin layer, whose thickness is usually related to the level of turbulence very close to the interface.

2.1. Film theory

The simplest model to quantify the interfacial gas exchange is the film theory [1]. In this model, gas transfer occurs through a thin "stagnant" film of laminar fluid adjacent to the air-water interface, and its thickness can be denoted as δ. According to the Fick's diffusion law, the gas transfer velocity can be determined as,

$$k = \frac{D}{\delta} \tag{2}$$

where D is molecular diffusivity of the dissolved gas.

In reality, a "stagnant" diffusivity boundary layer (DBL) with a constant thickness is a poorly modeled concept, actual DBL is disrupted by sporadic "bursts" of turbulent "eddies", which are large scale turbulent motions that lift "fresh" fluids to the surface and create a renewed exposure to the air phase [2, 3]. The gas transfer is still driven by molecular diffusion during these exposures. Therefore, the exposure time θ_s becomes a key parameter to control the flux across the interface. And the thickness of near surface DBL is a dynamic value, which is related to properties of impinging turbulent eddies.

2.2. Penetration model and surface renewal model

In the penetration model [2], near surface DBL is periodically disrupted by penetrating eddies from the bulk water body with a constant exposure time. In contrast, the surface renewal model [3] considered the renewal time as a random variable with an exponential probability distribution. Accordingly, the averaged gas transfer velocity can be modelled as

$$k = \sqrt{\frac{4D}{\pi \theta_s}} \quad \text{(Penetration model)} \tag{3}$$

$$k = \sqrt{\frac{D}{\theta_s}} \text{ or } k = \sqrt{Df} \quad \text{(Surface renewal model)} \tag{4}$$

where f is the surface renewal frequency. In these two models, gas transfer velocity is related to the diffusivity as

$$k \sim D^{1/2} \tag{5}$$

Both laboratory and fields studies have shown that k is better modeled by the surface renewal model than by the "stagnant film" model [4, 5]. And k is observed to be $k \sim D^n$, where n varies between 1/2 and 2/3, depending on free surface roughness and near surface hydrodynamics [6].

2.3. Random eddy model

Differed slightly from the concept of surface renewal, Harriott [7] proposed that near surface random eddies would also enhance the gas flux while approaching air-water interface even without completely renewing the interfacial layer. Laboratory study of renewal events of a thermal boundary layer (TBL) has proven that significant fraction of renewal events do not renew the TBL completely [8]. Gas transfer is therefore controlled by the eddy penetration depth and the lifetime of those random eddies [9].

Many efforts have been made to parameterize the average time interval between surface renewals through the properties of near surface random eddies. The "large eddy model" [10] argued that the renewal time scale is scaled with the largest turbulent eddies, which suggested,

$$\theta_s \sim L / u' \tag{6}$$

where L is the integral length scale and u' is the root-mean-square of fluctuating turbulent velocities.

On the other hand, the "small eddy model" [11, 12] suggested that the smallest eddies are the controlling mechanism of interfacial gas exchange. Thus the renewal time scale is determined by the Kolmogorov time scale,

$$\theta_s \sim (v / \varepsilon)^{1/2} \tag{7}$$

where v is kinematic viscosity and ε is turbulent dissipation rate at the interface.

If we substitute the renewal time scale into the surface renewal model (e.g. equation (4)), k can be written as,

$$k \sim Sc^{-1/2}u'Re_t^{-1/2} \; \text{(large eddy model)} \tag{8}$$

$$k \sim Sc^{-1/2}u'Re_t^{-1/4} \; \text{(small eddy model)} \tag{9}$$

where Sc is Schmidt number defined as $Sc = v / D$, Re_t is turbulent Reynolds number, which is defined as,

$$Re_t = u'L/v. \tag{10}$$

In the "small scale eddy" model, dissipation rate has been scaled with the large scale eddies as,

$$\varepsilon \sim u'^3 / L, \tag{11}$$

following the concept of turbulent energy cascade. These hydrodynamic models agreed well with both laboratory and field measurements on the interfacial gas transfer velocity. Chu and Jirka [13] conducted simultaneous measurements on turbulence and gas concentration in a grid-stirred tank to reveal the relation between large eddy motions and gas transfer process with the turbulent Reynolds number varying from 80 to 660. Small scale eddy motions have also been shown to be correlated with gas flux in a variety of experiments [14-16]. In the "small scale eddy" model, gas transfer velocity is generally expressed explicitly as related with the near surface turbulence dissipation rate,

$$k \sim Sc^{-1/2}(v\varepsilon)^{1/4} \tag{12}$$

The "two regime model" proposed by Thoefanus et al. [17] combined the "large eddy model" and the "small eddy model" by arguing that different size of the near surface turbulent eddies dominate interfacial gas flux mechanism depending on the turbulent Reynolds number. That is, the "large eddy model" is more appropriate at low turbulent Reynolds numbers ($Re_t < 500$), and small scale eddies are more relevant to high turbulent Reynolds number flows ($Re_t > 500$),

$$k = 0.73Sc^{-1/2}u'Re_t^{-1/2} \; \text{at } Re_t < 500 \tag{13}$$

$$k = 0.25Sc^{-1/2}u'Re_t^{-1/4} \; \text{at } Re_t > 500 \tag{14}$$

2.4. Surface divergence model

Based on the source layer theory (or blocking theory) and considering the transport of homogeneous and isotropic turbulence in the far field away from the free surface without tangential shear [18], Banerjee [19] provided a "surface divergence model" that relates the gas transfer velocity to the divergence of horizontal velocities on the air-water interface,

$$k \sim Sc^{-1/2} u' \mathrm{Re}_t^{-1/2} \left[\left\langle \frac{\partial u'}{\partial x} + \frac{\partial v'}{\partial y} \right\rangle^2 \right]_{int}^{1/4} \tag{15}$$

where u' and v' are fluctuating horizontal velocities, and subscript "int" denotes the air-water interface.

Banerjee, Lakehal [20] pointed out that the "surface divergence" physically is the signature of turbulent "sweep" events representing local "upwelling" motions at the surface from the bulk fluid. On the other hand, it is an alternative expression of the surface renewal and more specifically modeled by tangential components of velocities at the interface. Csanady [21] emphasized the role of breaking wavelet at the interface with high surface divergence that squeezes the DBL by "upwelling" motions, i.e., the disruption of DBL by turbulent eddies is enhanced during micro-wave breaking events. Recently, surface divergence has been shown to correlate with interfacial gas transfer process in laboratory studies [22, 23].

One advantage of the "surface divergence" model is that it can be easily implemented: (1) concept of "surface divergence" replaces the renewal time scale by velocity fluctuating motions, while the renewal time varies due to different environmental flow conditions (2) "surface divergence" is easier to be measured than the renewal time (e.g. through the surface PIV measurement using infrared imaging techniques at the water surface [23]).

3. Measurement techniques on interfacial gas transfer

Considering the fact that the interfacial gas transfer is ultimately limited by the very thin layer of the DBL (on the order of micrometers), the existent measurement technologies are hardly directly measuring the gas transfer velocity across the air-water interface. Most applied measurement techniques are indirect methods, e.g., measuring a designed tracer flux across the air-water interface then convert it to the gas of interest assuming that both are controlled by the same near surface turbulence then the transfer velocity is scaled by the molecular diffusivity (i.e. equation (5) in the surface renewal model).

3.1. Deliberate volatile trace experiments

Inert volatile tracers have been widely used in determining gas transfer velocities in field studies through a mass balance approach. For example, sulfur hexafluoride (SF_6) was deliberately added to water bodies to quantify the gas transfer velocity as a function of wind speed [5, 24-27], since it can be detected at a very low level in water with an excellent signal-to-noise ratio. Based on mass balance approach, gas transfer velocity can be determined,

$$k \approx \frac{h}{t_2 - t_1} \ln \frac{C_{wt_1}}{C_{wt_2}} \tag{16}$$

where C_{wt} is the concentration of released tracer in water at time t. h is the mean depth of the mixed layer. Originally, the tracer experiment is designed for closed lake with relatively small size [27]. The experiment time scale is on the order of days to weeks depending on the size of lakes. However, for a large lake or ocean, the concentration of tracer decreases quickly due to horizontal transport and dispersion. Meanwhile, the mixing layer depth may vary significantly in space as the surface area and volume exposed to the atmosphere increases due to dispersion effect.

The tracer method can be improved by co-releasing a second inert tracer with a different diffusion coefficient (e.g. ^{3}He). By releasing two tracers with a constant ratio, the decreases of concentration due to dispersion are the same for the two gases, but different due to interfacial exchange. Since we know the transfer velocity should differ by a factor of 3, as $Sc(^3He)$ is about eight times smaller than $Sc(SF_6)$, the effect of horizontal dispersion can be separated out. The dual tracer technique has been used to measure gas exchange in different water bodies [28-31]. Besides the dual tracer technique, a third nonvolatile tracer (e.g. bacterial spores and rhodamines) can also be introduced to determine the gas transfer velocity independently based on an arbitrary tracer pair [32].

3.2. Active controlled flux technique (proxy technique)

The active controlled flux technique (ACFT) is a method to quantify the gas transfer velocity through the analogy with the heat transfer across the air-water interface [33, 34]. One example is to use an infrared laser to heat a certain area of water surface. A sensitive infrared imager is used to capture the time series of images of the heated patch on water surface. In order to determine the renewal frequency f, the "surface renewal model" is employed to fit the observed average surface temperature decay curve. The transfer rate of heat can be calculated as,

$$k_H = \sqrt{D_H f} \tag{17}$$

where D_H is thermal conductivity of water. Thus, gas transfer velocity can be estimated as,

$$k_G = k_H \left(\frac{Sc}{Pr} \right)^{-n} \tag{18}$$

where Sc is Schmidt number of gas of interest and Pr is Prandtl number defined as $Pr = v / D_H$, the exponent n varies in the range between 1/2 and 2/3 depending on the roughness of water surface [6, 35]. Using this technique, Garbe, Schimpf [36] have experimentally demonstrated the probability density function of the surface renewal time can be described with a lognormal distribution.

However, several experiments found that discrepancies exist between estimates of transfer velocity based on ACFT and that from dual tracers measurements [9] or direct covariance method [37, 38]. Atmane, Asher [9] found gas transfer velocity (as referenced to $Sc = 600$) determined by ACFT (using heat as proxy) was overestimated by a factor of 2, approximately.

The discrepancy can be attributed to the fact that the random eddies might not take effect on heat and gas exchange equally through renewal events. The Sc number is typically much greater (e.g. $Sc(CO_2)$ is 600 at 25°C in fresh water) than the Pr (e.g. Pr is around 7 at 20°C in water) number, hence the thickness of the gas DBL is significantly smaller than that of the TBL. Some of the "upwelling" eddies might not approach the gas DBL but they can disturb the TBL effectively. Asher, Jessup [39] proposed a different scaling with Sc number and provided a solution using the surface penetration model. Atmane, Asher [9] argued that the eddy approaching distance needs to be included as an extension to the surface renewal model.

3.3. Eddy covariance (correlation) method

The vertical flux of a scalar of interest (e.g. temperature, moisture, CO_2 concentration) can be estimated by evaluating the covariance between the fluctuating vertical velocity component and the fluctuating scalar concentration measured simultaneously at a certain height above the air-water interface. With the method, horizontal homogeneity is assumed and Reynolds decomposition is applied. The "eddy flux" is written as,

$$F = \overline{w'c'} \tag{19}$$

In order to apply the eddy covariance method, fast response instrumentations are required to capture the high frequency fluctuations of the gas concentration and the turbulent velocity, if we intent to measure gas transfer across the air-water interface. The eddy covariance method has been applied to measure the air-sea CO_2 flux from the air side [37, 40] and DO (Dissolved Oxygen) flux from the aqueous side [13]. Applying eddy covariance method from the air side on the open ocean can be challenging due to the contamination of flow induced by the movement of ship-based platform and the uncertainty of gas concentration due to changes in air density caused by variations of temperature and water vapor known as the Webb effect [41].

Alternative to the eddy covariance method, a relaxed eddy accumulation (EA) method [42-44] was developed and employed to estimate the gas flux by separating measurement of gas concentration from updrafts and downdrafts. This method avoids the requirement of high frequency measurement on the fluctuating gas concentration.

Recently, measurements of turbulent flux with particle image velocimetry (PIV) and laser induced fluorescence (LIF) techniques [45] were conducted in a grid stirred tank. Herlina and Jirka [46] suggested that the gas transfer at different turbulent levels can be associated with different dominant eddy sizes according to the spectra of covariance terms, which agreed with the "two regime" theory [17].

In the field, eddy covariance method has been widely applied to measure DO flux across the water-sediment interface [47, 48]. Recently, a waterside direct covariance measurement [49] has also been conducted in the field to investigate the air-sea gas exchange under extreme wind speed conditions. Although the requirement of high sampling rate can be relaxed due to longer time and length scales of turbulence on the waterside than the air side, the isotropic turbulence assumption still needs to be invoked and justified.

3.4. Floating chamber measurements

Gas flux across the air-water interface can also be estimated by monitoring the change of gas concentration in the floating chamber (FC) [50] due to interface gas exchange over a certain period of time. Kremer, Nixon [51] suggested that FC method would be applicable for low to moderate wind conditions (less than 8-10 m/s) and with a limited fetch such that waves are young and nonbreaking. An ideal chamber should have a large ratio of water surface area to chamber volume. Matthews, St Louis [52] compared the CO_2 and CH_4 fluxes based on the FC method, tracer technique and wind dependence estimation. The result showed that the FC method overestimate the transfer velocity in low wind shear condition. Guerin, Abril [53] conducted FC measurements in reservoirs and rivers, which gave similar results with the eddy covariance technique. FC method was also applied in coastal regions [54] under low to moderate wind (<10 m/s) and weak current condition (<20 cm/s). The results showed overestimation on transfer velocity compared with wind dependent relationship. Vachon, Prairie [55] tested the FC method with dissipation rate measurement. The results showed that the artificial effect of FC on near surface turbulence depends strongly on the background turbulence level, that is, overestimation by FC method is relatively large in a low turbulence environment.

4. Driving forces and parameterizations

4.1. wind speed

Most experimental work and modeling on gas transfer velocity are based on wind speed measurements and parameterization. Although it is not a direct driving force on interfacial gas transfer, wind stress has been considered as the primary source of near surface turbulence. Overall, wind speed is a reliable parameterization variable and is found to agree well with experimental data on gas transfer velocity. The advantage of wind speed models is that wind speed can be easily measured or obtained through meteorological modeling or remote sensing thus it can be easily implemented into regional and global gas flux estimations.

Although it is difficult to measure the wind speed with the accuracy that is required for modelling the gas transfer velocity, [32, 56-58], a large amount laboratory and field experiments [26, 32, 56, 59-64] have been conducted to estimate the empirical relationship between wind speed and gas transfer velocity and they are summarized in the following.

The first wind speed model was presented by Liss and Merlivat [61]. A "three linear segments" relationship between the gas transfer velocity and the wind speed was proposed based on wind tunnel experiments. The three segments were categorized according to the surface roughness (smooth surface, $U_{10} < 3.6$ m/s; rough surface, 3.6 m/s $< U_{10} < 13$ m/s; breaking wave region, $U_{10} > 13$ m/s).

Up to the present day, the most popular wind speed based gas transfer model is a quadratic relation. Wanninkhof [62] suggested that gas transfer velocity scales with U_{10}^2 [26] based on the global bomb ^{14}C constraint [65] and wind wave tank results. The quadratic relation indicates that the gas transfer scales with wind stress as $\tau \sim C_D U_{10}^2$. The quadratic relationship [62] for gas transfer velocity of CO_2 at 20°C for seawater ($Sc = 660$) is written as

$$k_{660} = 0.39 \langle U_{10} \rangle^2 \qquad (20)$$

where the transfer velocity is expressed in "cm/hour" and wind speed is in "m/s". Furthermore, Wanninkhof [62] modified the scaling factor for the cases of short-term or steady wind conditions,

$$k_{660} = 0.31 \langle U_{10} \rangle^2 \qquad (21)$$

Similar quadratic relationship was derived by Nightingale, et al. [32] from deliberate tracer experiments in the coastal ocean:

$$k_{660} = 0.222 U_{10}^2 + 0.333 U_{10} \qquad (22)$$

This result is in between the model of Liss and Merlivat [61] and that of Wanninkhof [62]. Recently, the SOLAS Air-Sea Gas Exchange (SAGE) experiment was conducted in the Southern Ocean [66]. The new quadratic relationship is given from dual tracer injection experiments as

$$k_{660} = (0.266 \pm 0.019) U_{10}^2 \qquad (23)$$

More recently, additional dual tracer experiments were conducted in Southern Ocean [67]. From the new data, the relationship [31] was updated to,

$$k_{660} = (0.262 \pm 0.022) U_{10}^2 \qquad (24)$$

Alternatively, a cubic relation was proposed by Wanninkhof and McGillis [64] for steady or short term wind conditions,

$$k_{660} = 0.0283U_{10}^3 \tag{25}$$

This relation is in good agreement with direct covariance results of air-sea Gas Exchange Experiment conducted in 1998 (GasEx-98). The cubic relation is supported by GasEx-98 data [37] and GasEx-2001 data [38] in the following expressions,

$$k_{660} = 0.026U_{10}^3 + 3.3 \tag{26}$$

$$k_{660} = 0.014U_{10}^3 + 8.2 \tag{27}$$

Although wind speed parameterization is probably the most convenient and a successful model [68] for estimating interfacial gas transfer velocity, the method is largely empirical. Most supporting data came from local experiments, which could be affected by many factors (such as the experiment location, measurement techniques, instrumentation errors and experimental uncertainties). Ho, Law [66] argued that the experiments of Nightingal, et al. [32] might be affected by an underdeveloped wind field and higher concentration of surfactants in coastal area. And the result of Wanninkhof [62] is most likely an overestimate because of an excessive ^{14}C inventory of the global ocean.

If the wind speed model were to apply to estimate the global CO_2 uptake by oceans, the global wind speed estimate would be a very critical issue. The total fluxes estimation is very sensitive to the accuracy of global wind speed estimation [69, 70]. Wanninkhof, Asher [71] pointed out since the long term averaged transfer velocity essentially scales with the second or third order of moment of the wind speed, the quadratic relationship gives a 27% higher result compared with the short term estimation while the cubic relationship gives a 91% higher result. It should be noted that the global wind speed distribution can be approximately represented by a Rayleigh distribution [62, 72].

4.2. Wind stress

In general, the relation between gas transfer velocity and wind speed can be summarized as,

$$k \sim Sc^{-n}U_{10}^b \tag{28}$$

where $b = 1, 2, 3$, representing linear, quadratic, cubic relations with respect to the wind speed. According to Charnock's Law [73],

$$\frac{U(z)}{u_{*_a}} = \kappa^{-1} \ln\left(\frac{gz}{u_{*_a}^2}\right) + C \tag{29}$$

where κ is von Kármán's constant. Meanwhile the surface shear stress caused by wind can be related to the wind speed as,

$$\tau = \rho_a C_D U_{10}^2 \tag{30}$$

where C_D is the wind drag coefficient, which is also a function of U_{10} [74, 75]. If we apply the continuity of shearing stress at the interface,

$$\tau = \rho_a u_{*_a}^2 = \rho_w u_{*_w}^2 \tag{31}$$

So the relation among the wind speed and friction velocities of the air and water sides can be,

$$U_{10} \sim u_{*_a}^\alpha \sim u_{*_w}^\alpha \tag{32}$$

where α depends on the scaling of the drag coefficient with the wind speed. Many experimental results suggested that the drag coefficient increases linearly with wind speed except for the case of low wind speed, so $\alpha = 1/2$ can be derived [74-78]. Also since the quadratic law is the most widely accepted wind speed model for gas transfer velocity, i.e., $b = 2$ in equation (28), the gas transfer velocity is linearly scaled with the water side shear velocity,

$$k \sim Sc^{-n} u_* \tag{33}$$

It's noting that for most wind speed models, the power of Schmidt number $-n$ is set to be $-1/2$, we have,

$$k \sim \sqrt{D} \tag{34}$$

which is consistent with the surface renewal model (equation (4)).

On the other hand, Jähne and Haußecker [35] shows that the gas transfer velocity can be expressed explicitly by interfacial shear velocity though turbulent diffusive boundary layer theory:

$$k \sim u_* Sc_t \tag{35}$$

where Sc_t is the turbulent Schmidt number, defined as the ratio of turbulent diffusion coefficient of momentum and gas concentration:

$$Sc_t = \frac{K_m}{K_c}$$

(36)

Using the concept of diffusive boundary layer Deacon [79] proposed that,

$$k = 0.082Sc^{-2/3}u_{*w}$$

(37)

which shows that gas transfer velocity is proportional to interfacial shear velocity.

The relation derived from the diffusive boundary layer theory is also similar to equation (33), which is derived from the empirical wind speed model (quadratic relation). The difference is the exponent of the Schmidt number. In Deacon [79]'s model, the -2/3 power scaling is suitable for smooth surface, as it is pointed out by Jähne and Haußecker [35]. The -1/2 power scaling is more appropriate for a wave-covered water surface [6]. Fairaill et al. [80] conducted a comprehensive analysis on a number of parameters including effects of shear forcing, roughness Reynolds number and buoyancy effects on the gas transfer. Their results have been applied by Hare et al. [81] to evaluate the GasEx data. And they found significant gas flux occurs due to wave breaking and air bubble entrainment, which will be discussed in the next section.

4.3. The effect of sea surface roughness, wave breaking and entranced air bubbles

Experiments confirmed that gas transfer is enhanced by the presence of wind induced ripples. From the perspective of momentum transport, turbulence can be enhanced by the increase of surface roughness. The exponent of the Schmidt number in wind speed models or wind shear models varies from about -2/3 to -1/2, which was found to be dependent on the surface roughness. For CO_2, that implies a variation in the transfer velocity by a factor of 3. Jähne et al. [6] demonstrated a good correlation between the gas transfer velocity and the mean square slope of surface waves in a wind/wave facility. Frew et al. [82]'s field experiments showed stronger correlation between the transfer velocity and the mean square slope compared to wind speed relation. Since the wave slope can be obtained through satellite-base remote sensing, this relation provides a method that can be easily implemented to estimate the global gas flux [83].

The majority of laboratory and field experiments on gas transfer were conducted under weak to moderate wind conditions. Extremely high wind speed makes the measurement very difficult. From the few existing data, the transfer velocity is significantly enhanced in high wind fields. The accepted theory is that the gas flux across the interface is dominated by wave breaking and entrained air bubbles [60, 84-88]. Woolf and Thorpe [89] argued that the transfer velocity is only enhanced by bubbles for very low soluble gases. Woolf [90] introduced a transfer velocity term which is specifically due to bubbles. Thus the transfer velocity can be expressed by a hybrid model [91],

$$k = k_0 + k_b \tag{38}$$

where k_b is approximately proportional to the whitecap coverage [92]. Factors that influence bubble mediated transfer were reviewed by Woolf [92] and Woolf et al. [93]. Alternatively, other parameterizations and analyses on gas transfer velocity through the whitecap coverage exist [94, 95].

4.4. The effect of surfactants

The presence of surfactants is believed to have an attenuation effect on interfacial gas exchange. Early laboratory experiments observed a large amount of reduction of transfer velocity due to the presence of surfactants [96]. Asher [97] reported a linear relationship between the transfer velocity and wind speed at the presence of surfactant when wind speed is smaller than 12.5 m/s.

Numerous studies of the effects of surfactants on air-sea gas transfer have been conducted in laboratory settings and *in situ* [14, 22, 98-102]. It's noting that some of surfactants are soluble, while others are not. The insoluble surfactant acts as a barrier film. However, this effect can be easily dispersed by wind and waves. For high wind condition, the soluble surfactants are believed to have a prevailing effect on gas transfer even at the presence of breaking waves, while insoluble surfactants do not [103-105].

4.5. The effect of rainfall

Air-sea gas exchange during rainfall events has been brought into attention recently. It has been shown that rainfall will enhance the transfer velocity across the interface [56, 106-109]. Existing evidence shows that the enhancement is due to rainfall generated turbulence and bubble entrainments. The kinetic energy flux (KEF) caused by raindrops has been introduced to scale with the gas transfer velocity [106, 110]. However, Takagaki and Komori [111] argued that transfer velocity is more correlated with the momentum flux of rainfall (MF).

The effect of raindrops on the enhancement of surface mixing, damping waves and changing the air-sea momentum flux has been investigated through the surface renewal model [112]. Rainfall could also induce surface density stratification and additional surface heat flux because of temperature difference between raindrops and the sea surface. The combined effect of rainfall and high wind speed is believed to have a significant impact on air-water gas exchange, however, this effect is complex and yet to be investigated comprehensively.

4.6. Near surface turbulence

The parameterizations of interfacial gas exchange discussed above are generally empirical or semi-empirical. For most empirical models, gas transfer velocities are scaled with meteorological parameters such as wind speed, wind shear, momentum flux or kinetic energy flux induced by rainfalls (rainfall dominant environment), etc. In comparison, models based on near surface turbulence structures, such as the surface renewal model and the surface divergence model, are more mechanistic. For gases with low solubility, the resistance of gas transfer is dominated by the water side, which is in turn controlled by the near surface turbulence.

A large amount experiments were conducted to investigate the near surface turbulence and its relation to air-sea gas transfer process. Lamont and Scott [11] presented an eddy cell model to quantify the mass transfer from the hydrodynamic parameters (equation (12)). Some recent studies show the gas transfer velocity is better scaled with the surface turbulence [14, 16, 109]. Zappa, McGillis [15] has shown that gas transfer velocity is well correlated with the dissipation rate rather than wind speed under a variety of environmental forcing, regardless the how the near surface turbulence was produced. Vachon et al. [55] performed a number of measurements to demonstrate the direct relationship between gas transfer velocity (measured by a floating chamber) and near surface turbulent dissipation rate (measured by an ADV). Lorke and Peeters [113] demonstrated that equation (12) can be derived by assuming the thickness of diffusive sub-layer to be scaled with the Batchelor's micro-scale,

$$\delta_D = L_B = 2\pi \left(\frac{\nu D^2}{\varepsilon} \right)^{1/4} \tag{39}$$

It is worth noting that the dissipation rate scaling is based on the assumption of homogenous and isotropic turbulence near the water surface. Correspondingly, the small eddy model is applicable for a high Reynolds number condition, which is the prerequisite of Kolmogorov's similarity hypothesis.

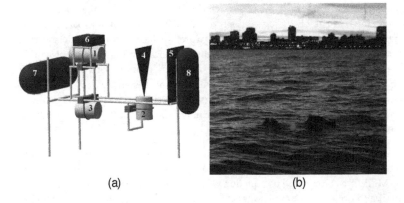

(a)   (b)

Figure 1. (a) The free floating UWMPIV. Components: (1) camera housing (2) laser housing (3) battery housing (4) laser sheet (5) guide plate (6) wireless router (7)(8) floating buoys (b) Deployment of UWMPIV on Lake Michigan

Despite the promising results from small scale eddy models, the scaling coefficient has not been clearly determined yet. In most of these studies, the coefficient is usually obtained empirically from fitting modeled transfer velocities with measured ones. Another uncertainty is associated with the depth at which the dissipation rate should be applied in the model. In theory the dissipation rate should be measured immediately below the air-water interface, but this cannot be easily achieved during field measurements. While there is very few *in situ* data available for the near surface turbulence, existing field data was obtained at a short distance (tenth of centimeters) away from the wavy surface. Zappa et al. [15]argued that this might not be a serious issue, as the gas transfer velocity should scale with $\varepsilon^{1/4}$. However, if a strong gradient of dissipation rate exists near the surface, this assumption has to be examined carefully.

Wang et al. [114] have recently developed a free floating Under Water Miniature Particle Velocimetry system (UWMPIV) (Figure 1) to measure vertical profiles of the turbulence dissipation rate immediately below the water surface, and has successfully deployed it on Lake Michigan. Figure 2 shows a sample of measured near surface turbulence structure. In order to calculate the vertical dissipation rate profile, velocity maps were evaluated on a dynamic triangular mesh with the moving air-water interface as the top boundary (see figure 2(a)). Statistics were obtained with the vertical coordinates attached to the local water surface where $z = 0$ and z increases with water depth. Figure 3 shows measured dissipation rate profiles under varying wind speed (U_{10} ranged from 2 to 15 m s^{-1}) and wave conditions. Runs 1~5 were measured in a harbor with essentially zero wind fetch and very short waves, whereas run 6 was measured on the open lake with weak wind ($U_{10} \approx 2$ m s^{-1}) and developed wave field (significant wave height ≈ 0.35 m). The detailed description of each runs can be found in Wang et al. [114].

A wide range of dissipation rate (from 10^{-6} to 10^{-3} m^{2}s^{-3}) was covered in the data series. From the case of run 6, it shows that surface waves might also be a significant source of surface turbulence since the wind speed is rather small in this case while the dissipation rate is comparable to that from the cases where wind speed was in the range of 10~15 m s^{-1} and a nearly zero fetch (run1-5. For all cases, a strong vertical gradient of dissipation rate was found, with peak dissipation immediately below the water surface, and then it decays rapidly with depth, usually by one order of magnitude within several centimeters. Profiles of dissipation rates can be described by a power law with the exponent ranging between 1 and 2. These new findings suggest that measurements of turbulence at some distance away from the surface may not be directly applied to estimate the gas transfer velocity at the surface. It also suggests that more efforts are needed to reveal the exact structure of small scale turbulence within several centimeters of the surface water.

Figure 4 compares estimated transfer velocity of CO_2 across the air-sea interface at 20°C seawater (Sc = 660) from three wind speed models and the small scale eddy model based on the UWMPIV measurement, i.e., equation (12) with the scaling factor = 0.419 following [15]. The dissipation rate in the small scale eddy model was measured at $z = 1$ cm below the air-water interface for all runs. For three wind speed models, W92 represents the short term or steady wind speed condition estimation [62]; N00 represents the coastal area measurement in fetch limited environments [32]; H06 relationship is obtained from SOLAS Air-Sea Gas Exchange (SAGE) experiment in Southern Ocean [66]. Since runs 1~5 were conducted in the Milwaukee Harbor with an almost "zero-fetch" condition, the dissipation rate can be considered as a representative of wind shear without any significant wave effects. Transfer velocity estimation based on dissipation rate match all wind speed models very well under conditions of moderate wind speed (5-10 m s^{-1}). For the case when the wind speed was about 15 m s^{-1}. The small scale eddy model might underestimate the actual gas transfer rate as significant bubble entrainments were observed for that case.

For run 6 under a low wind condition, wind speed modeled transfer velocity decreases, however the near surface turbulent dissipation is enhanced possibly due to the interaction among non-breaking waves or the micro-breaking events. The transfer velocity estimated by the small scale eddy model was almost 10 times larger than that of wind speed models. This observation suggested surface waves themselves might contribute to produce the near surface turbulence. Therefore the wind speed model may have significantly underestimated the global air-sea gas transfer, since the sea surface is subject to low to moderate wind speed for most of the time [115], while surface waves are present almost all the time.

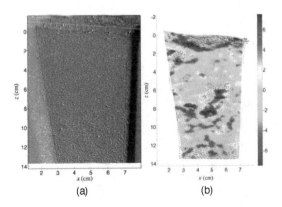

Figure 2. (a) A sample image pair with triangular PIV mesh (b) the instantaneous velocity vector map superimposed on the calculated vorticity map, the unit of the color bar is (s^{-1}) [114]

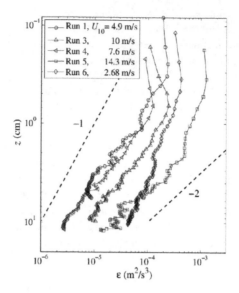

Figure 3. The dissipation rate profiles at different wind shear and wave conditions (log-log scale) [114]

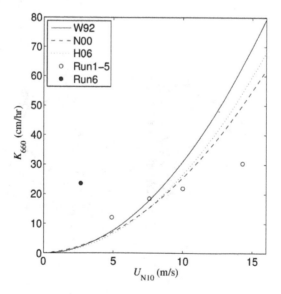

Figure 4. Comparison of CO$_2$ transfer velocity models based on wind speed and the surface turbulence dissipation rate.

5. Conclusion

Gas transfer across the atmosphere/ocean interface is a very important physical process that regulates the global climate, considering the fact that this process occurs over an area that is about 70% of the Earth surface. This chapter provides a review on the current technologies of measuring the gas flux across the air/ocean surface and existing models for the gas transfer velocity. Many environmentally important gases, such as O_2 or CO_2, have a low solubility, so the major resistance of gas exchange is from the water side. Near surface turbulence might be the key physical parameter that determines the gas transfer velocity as it controls the thickness of the diffusive boundary layer, e.g., "eddy" upwelling induced surface renewal. Global gas flux estimates still use the wind speed, or wind shear stress as the primary modeling parameter, as wind is the major source of near surface turbulence. For example W92 model is commonly applied to estimate the global air-sea CO_2 flux. Additionally, breaking wave parameters such as the whitecap coverage, are also included to account for gas exchange through bubbles induced by breaking waves. Other environmental forcing may also be significant sources of near surface turbulence that can affect the gas flux. Turbulence can be generated by bottom mixing then diffuses to the surface in a coastal area; turbulent buoyant convection, surface currents, precipitation and microbreaking of short surface waves can also contribute to near surface turbulence production. These parameters are usually not directly related to the wind stress.

Recent studies indicated that small scale eddy models parameterized with the surface dissipation rate or divergence are more mechanistic thus a more universal approach to estimate the gas transfer velocity under a wide range of environmental forcing conditions, except for the case of breaking waves. Advances in measurement technologies, such as the floating UWMPIV [114], provide encouraging opportunities to quantify the structure of turbulence in the upmost layer below the air-sea interface. Preliminary results showed that surface wave itself might be a source of near surface turbulence and it can significantly enhance the gas transfer velocity under low wind conditions. Future research on this subject should focus on the scaling of the surface turbulence structure with properties of large scale environmental forcing. Simultaneous field measurements of the flow field and the gas transfer velocity are also needed to provide further insights into air-sea gas exchange processes.

Acknowledgements

Part of the research presented here was supported by the US National Science Foundation under Grant No. 0826477, and the Wisconsin Sea Grant under projects R/HCE-3 and R/HCE-11.

Author details

Qian Liao* and Binbin Wang

Department of Civil Engineering and Mechanics, University of Wisconsin-Milwaukee, Wisconsin, USA

References

[1] Lewis, W. K, & Whitman, W. G. Principles of Gas Absorption. Ind Eng Chem. (1924). , 16(12), 1215-20.

[2] Higbie, R. The rate of absorption of a pure gas into a still liquid during short periods of exposure. AIChE Transactions. (1935). , 31, 365-90.

[3] Danckwerts, P. V. Sinificance of liquid-film coefficients in gas absorption. Ind Eng Chem. (1951). , 43, 1460-7.

[4] Jähne, B, Huber, W, Dutzi, A, Wais, T, & Ilmberger, J. Wind/wave-tunnel experiments on the Schmidt number and wave field dependence of air-water gas exchange. In: Brutsaert W, Jirka GH, editors. Gas transfer at water surfaces(1984). , 303-309.

[5] Upstill-goddard, R. C, Watson, A. J, Liss, P. S, & Liddicoat, M. I. Gas transfer velocities in lakes measured with SF_6. Tellus B. (1990). , 42, 364-77.

[6] Jähne, B, Munnich, K. O, Bosinger, R, Dutzi, A, Huber, W, & Libner, P. On the Parameters Influencing Air-Water Gas-Exchange. J Geophys Res-Oceans. (1987). C2): 1937-49.

[7] Harriott, P. A random eddy modification of the penetration theory. Chem Eng Sci. (1962). , 17, 149-54.

[8] Jessup, A. T, Asher, W. E, Atmane, M, Phadnis, K, Zappa, C. J, & Loewen, M. R. Evidence for complete and partial surface renewal at an air-water interface. Geophys Res Lett. (2009).

[9] Atmane, M. A, Asher, W. E, & Jessup, A. T. On the use of the active infrared technique to infer heat and gas transfer velocities at the air-water free surface. J Geophys Res-Oceans. (2004). C8).

[10] Fortescue, G. E. Pearson JRA. On gas absorption into a turbulent liquid. Chem Eng Sci. (1967). , 22, 1163-76.

[11] Lamont, J. C, & Scott, D. S. An eddy cell model of mass transfer into the surface of a turbulent liquid. Aiche J. (1970). , 16, 513-9.

[12] Banerjee, S, Rhodes, E, & Scott, D. S. Mass transfer through falling wavy liquid films in turbulent flow. Industrial and Engineering Chemicals Fundamentals. (1968). , 7, 22-7.

[13] Chu, C. R, & Jirka, G. H. Turbulent Gas Flux Measurements Below the Air-Water-Interface of a Grid-Stirred Tank. Int J Heat Mass Tran. (1992). , 35(8), 1957-68.

[14] Asher, W. E, & Pankow, J. F. The interaction of mechanically generated turbulence and interfacial films with a liquid phase controlled gas/liquid transport process. Tellus B. (1986). , 38, 305-18.

[15] Zappa, C. J, Mcgillis, W. R, Raymond, P. A, Edson, J. B, Hintsa, E. J, Zemmelink, H. J, et al. Environmental turbulent mixing controls on air-water gas exchange in marine and aquatic systems. Geophys Res Lett. (2007).

[16] Zappa, C. J, Raymond, P. A, Terray, E. A, & Mcgillis, W. R. Variation in surface turbulence and the gas transfer velocity over a tidal cycle in a macro-tidal estuary. Estuaries. (2003). , 26(6), 1401-15.

[17] Theofanus, T. G, Houze, R. N, & Brumfield, L. K. Turbulent mass transfer at free, gas liquid interfaces with applications to open channel, bubble and jet flows. Int J Heat Mass Tran. (1976). , 19(6), 613-24.

[18] Hunt JCRGraham JMR. Free stream turbulence near plane boundaries. J Fluid Mech. (1978). , 84, 209-35.

[19] Banerjee, S. Turbulence structure and transport mechanisms at interfaces. Proc 9th International Heat Transfer Conference(1990). , 395-418.

[20] Banerjee, S, Lakehal, D, & Fulgosi, M. Surface divergence models for scalar exchange between turbulent streams. Int J Multiphas Flow. (2004).

[21] Csanady, G. T. The Role of Breaking Wavelets in Air-Sea Gas Transfer. J Geophys Res-Oceans. (1990). C1):749-59.

[22] Mckenna, S. P, & Mcgillis, W. R. The role of free-surface turbulence and surfactants in air-water gas transfer. Int J Heat Mass Tran. (2004). , 47(3), 539-53.

[23] Asher, W. E, Liang, H, Zappa, C. J, Loewen, M. R, Mukto, M. A, Litchendorf, T. M, et al. Statistics of surface divergence and their relation to air-water gas transfer velocity. J Geophys Res-Oceans. (2012). C05035).

[24] Wanninkhof, R, Ledwell, J. R, & Broecker, W. S. Gas exchange-wind speed relation measured with sulfur hexafluoride on a lake. Science. (1985). Epub 1985/03/08., 227(4691), 1224-6.

[25] Wanninkhof, R, Ledwell, J. R, Broecker, W. S, & Hamilton, M. Gas-Exchange on Mono Lake and Crowley Lake, California. J Geophys Res-Oceans. (1987). C13): 14567-80.

[26] Wanninkhof, R. H, & Bliven, L. F. Relationship between Gas-Exchange, Wind-Speed, and Radar Backscatter in a Large Wind-Wave Tank. J Geophys Res-Oceans. (1991). C2):2785-96.

[27] Clark, J. F, Schlosser, P, Wanninkhof, R, & Simpson, H. J. Schuster WSF, Ho DT. Gas Transfer Velocities for Sf6 and He-3 in a Small Pond at Low Wind Speeds. Geophys Res Lett. (1995). , 22(2), 93-6.

[28] Watson, A. J, Upstillgoddard, R. C, & Liss, P. S. Air Sea Gas-Exchange in Rough and Stormy Seas Measured by a Dual-Tracer Technique. Nature. (1991). , 349(6305), 145-7.

[29] Wanninkhof, R, Sullivan, K. F, & Top, Z. Air-sea gas transfer in the Southern Ocean. J Geophys Res-Oceans. (2004). C8).

[30] Ho, D. T, Law, C. S, Smith, M. J, Schlosser, P, Harvey, M, & Hill, P. Measurements of air-sea gas exchange at high wind speeds in the Southern Ocean: Implications for global parameterizations. Geophys Res Lett. (2006).

[31] Ho, D. T, Wanninkhof, R, Schlosser, P, Ullman, D. S, Hebert, D, & Sullivan, K. F. Toward a universal relationship between wind speed and gas exchange: Gas transfer velocities measured with (3)He/SF(6) during the Southern Ocean Gas Exchange Experiment. J Geophys Res-Oceans. (2011).

[32] Nightingale, P. D, Malin, G, Law, C. S, Watson, A. J, Liss, P. S, Liddicoat, M. I, et al. In situ evaluation of air-sea gas exchange parameterizations using novel conservative and volatile tracers. Global Biogeochem Cy. (2000). , 14(1), 373-87.

[33] Haußecker HJähne B. In situ measurements of the air-sea gas transfer rate during the MBL/CoOP west coast experiment. In: Jähne B, Monahan EC, editors. Air Water Gas Transfer(1995). , 775-784.

[34] Haußecker HJähne B, Reinelt S. Heat as a proxy tracer for gas exchange measurements in the field: principles and technical realization. In: Jähne B, Monahan EC, editors. Air Water Gas Transfer(1995). , 405-413.

[35] Jähne, B. Haußecker H. Air-water gas exchange. Annu Rev Fluid Mech. (1998). , 30, 443-68.

[36] Garbe, C. S, Schimpf, U, & Jähne, B. A surface renewal model to analyze infrared image sequences of the ocean surface for the study of air-sea heat and gas exchange. J Geophys Res-Oceans. (2004). C8).

[37] Mcgillis, W. R, Edson, J. B, Hare, J. E, & Fairall, C. W. Direct covariance air-sea CO2 fluxes. J Geophys Res-Oceans. (2001). C8):16729-45.

[38] Mcgillis, W. R, Edson, J. B, Zappa, C. J, Ware, J. D, Mckenna, S. P, Terray, E. A, et al. Air-sea CO2 exchange in the equatorial Pacific. J Geophys Res-Oceans. (2004). C8).

[39] Asher, W. E, Jessup, A. T, & Atmane, M. A. Oceanic application of the active controlled flux technique for measuring air-sea transfer velocities of heat and gases. J Geophys Res-Oceans. (2004). C8).

[40] Edson, J. B, Hinton, A. A, Prada, K. E, Hare, J. E, & Fairall, C. W. Direct covariance flux estimates from mobile platforms at sea. J Atmos Ocean Tech. (1998). , 15(2), 547-62.

[41] Webb, E. K, Pearman, G. I, & Leuning, R. Correction of flux measurements for density effects due to heat and water vapour transfer. Q J Roy Meteor Soc. (1980). , 106, 85-100.

[42] Businger, J. A, & Oncley, S. P. Flux Measurement with Conditional Sampling. J Atmos Ocean Tech. (1990)., 7(2), 349-52.

[43] Zemmelink, H. J. Gieskes WWC, Klaassen W, de Groot HW, de Baar HJW, Dacey JWH, et al. Simultaneous use of relaxed eddy accumulation and gradient flux techniques for the measurement of sea-to-air exchange of dimethyl sulphide. Atmos Environ. (2002).

[44] Zemmelink, H. J. Dacey JWH, Hintsa EJ. Direct measurements of biogenic dimethylsulphide fluxes from the oceans: a synthesis. Can J Fish Aquat Sci. (2004)., 61(5), 836-44.

[45] HerlinaJirka GH. Application of LIF to investigate gas transfer near the air-water interface in a grid-stirred tank. Exp Fluids. (2004)., 37(3), 341-9.

[46] HerlinaJirka GH. Experiments on gas transfer at the air-water interface induced by oscillating grid turbulence. J Fluid Mech. (2008)., 594, 183-208.

[47] Berg, P, Roy, H, Janssen, F, Meyer, V, Jorgensen, B. B, Huettel, M, et al. Oxygen uptake by aquatic sediments measured with a novel non-invasive eddy-correlation technique. Mar Ecol-Prog Ser. (2003)., 261, 75-83.

[48] Berg, P, Glud, R. N, Hume, A, Stahl, H, Oguri, K, Meyer, V, et al. Eddy correlation measurements of oxygen uptake in deep ocean sediments. Limnol Oceanogr-Meth. (2009)., 7, 576-84.

[49] Asaro, D, Mcneil, E, & Air-sea, C. gas exchange at extreme wind speeds measured by autonomous oceanographic floats. J Marine Syst. (2007).

[50] Frankignoulle, M. Field measurements of air-sea CO_2 exchange. Limnol Oceanogr. (1988)., 33(3), 313-22.

[51] Kremer, J. N, Nixon, S. W, Buckley, B, & Roques, P. Technical note: Conditions for using the floating chamber method to estimate air-water gas exchange. Estuaries. (2003). A):, 985-90.

[52] Matthews CJDSt Louis VL, Hesslein RH. Comparison of three techniques used to measure diffusive gas exchange from sheltered aquatic surfaces. Environ Sci Technol. (2003)., 37(4), 772-80.

[53] Guerin, F, Abril, G, Serca, D, Delon, C, Richard, S, Delmas, R, et al. Gas transfer velocities of CO(2) and CH(4) in a tropical reservoir and its river downstream. J Marine Syst. (2007).

[54] Tokoro, T, Watanabe, A, Kayanne, H, Nadaoka, K, Tamura, H, Nozakid, K, et al. Measurement of air-water CO2 transfer at four coastal sites using a chamber method. J Marine Syst. (2007).

[55] Vachon, D, Prairie, Y. T, & Cole, J. J. The relationship between near-surface turbu-
     lence and gas transfer velocity in freshwater systems and its implications for floating
     chamber measurements of gas exchange. Limnol Oceanogr. (2010). , 55(4), 1723-32.

[56] Frost, T, & Upstill-goddard, R. C. Meteorological controls of gas exchange at a small
     English lake. Limnol Oceanogr. (2002). , 47(4), 1165-74.

[57] Upstill-goddard, R. C. Air-sea gas exchange in the coastal zone. Estuar Coast Shelf S.
     (2006). , 70(3), 388-404.

[58] Yelland, M. J, Moat, B. I, Taylor, P. K, Pascal, R. W, Hutchings, J, & Cornell, V. C.
     Wind stress measurements from the open ocean corrected for airflow distortion by
     the ship. J Phys Oceanogr. (1998). , 28(7), 1511-26.

[59] Liss, P. S. Gas transfer: experiments and geochemical implications. In: Liss PS, Slinn
     WG, editors. Air-sea exchange of gases and partilces(1983). , 241-298.

[60] Merlivat, L, & Memery, L. Gas exchange across an air-water interface: experimental
     results and modelling of bubble contribution to transfer. J Geophys Res-Oceans.
     (1983). , 88, 707-24.

[61] Liss, P. S, & Merlivat, L. Air-sea gas exchange rates: introduction and synthesis. In:
     Buat-Menard P, editor. The role of air-sea exchange in geochemical cycling(1986). ,
     113-129.

[62] Wanninkhof, R. Relationship between Wind-Speed and Gas-Exchange over the
     Ocean. J Geophys Res-Oceans. (1992). C5):7373-82.

[63] Wanninkhof, R, Asher, W, Weppernig, R, Chen, H, Schlosser, P, Langdon, C, et al.
     Gas Transfer Experiment on Georges Bank Using 2 Volatile Deliberate Tracers. J Geo-
     phys Res-Oceans. (1993). C11):20237-48.

[64] Wanninkhof, R, & Mcgillis, W. R. A cubic relationship between air-sea CO2 exchange
     and wind speed. Geophys Res Lett. (1999). , 26(13), 1889-92.

[65] Broecker, W. S, Peng, T. H, Östlund, G, & Stuiver, M. The distribution of bomb radio-
     carbon in the ocean. J Geophys Res-Oceans. (1985). , 99, 6953-70.

[66] Ho, D. T, Law, C. S, Smith, M. J, Schlosser, P, Harvey, M, & Hill, P. Measurements of
     air-sea gas exchange at high wind speeds in the Southern Ocean: Implications for
     global parameterizations (art Geophys Res Lett. (2006). , 33(L16611)

[67] Ho, D. T, Sabine, C. L, Hebert, D, Ullman, D. S, Wanninkhof, R, Hamme, R. C, et al.
     Southern Ocean Gas Exchange Experiment: Setting the stage. J Geophys Res-Oceans.
     (2011).

[68] Takahashi, T, Sutherland, S. C, Sweeney, C, Poisson, A, Metzl, N, Tilbrook, B, et al.
     Global sea-air CO2 flux based on climatological surface ocean pCO(2), and seasonal
     biological and temperature effects. Deep-Sea Res Pt Ii. (2002).

[69] Boutin, J, Etcheto, J, Merlivat, L, & Rangama, Y. Influence of gas exchange coefficient parameterisation on seasonal and regional variability of CO(2) air-sea fluxes. Geophys Res Lett. (2002).

[70] Naegler, T, Ciais, P, Rodgers, K, & Levin, I. Excess radiocarbon constraints on air-sea gas exchange and the uptake of CO2 by the oceans. Geophys Res Lett. (2006).

[71] Wanninkhof, R, Asher, W. E, Ho, D. T, Sweeney, C, & Mcgillis, W. R. Advances in Quantifying Air-Sea Gas Exchange and Environmental Forcing. Annu Rev Mar Sci. (2009). , 1, 213-44.

[72] Wentz, F. J, Peteherych, S, & Thomas, L. A. A model function for ocean radar cross section at 14.6 GHz. J Geophys Res-Oceans. (1984). , 83, 3689-704.

[73] Charnock, H. Wind stress on a water surface. Q J Roy Meteor Soc. (1955). , 81, 639-40.

[74] Yelland, M, & Taylor, P. K. Wind stress measurements from the open ocean. J Phys Oceanogr. (1996). , 26(4), 541-58.

[75] Large, W. G, & Pand, S. Open ocean momentum flux measurements in moderate to strong winds. J Phys Oceanogr. (1981). , 11, 324-36.

[76] Geernaert, G. L, Davidson, K. L, Larsen, S. E, & Mikkelsen, T. Wind Stress Measurements during the Tower Ocean Wave and Radar Dependence Experiment. J Geophys Res-Oceans. (1988). C11):13913-23.

[77] Smith, S. D, Anderson, R. J, Oost, W. A, Kraan, C, Maat, N, Decosmo, J, et al. Sea-Surface Wind Stress and Drag Coefficients- the Hexos Results. Bound-Lay Meteorol. (1992).

[78] Anderson, R. J. A Study of Wind Stress and Heat-Flux over the Open-Ocean by the Inertial-Dissipation Method. J Phys Oceanogr. (1993). , 23(10), 2153-61.

[79] Deacon, E. L. Gas transfer to and across an air-water interface. Tellus. (1977). , 29, 363-74.

[80] Fairaill, C. W, Hare, J. E, Edson, J. B, & Mcgillis, W. Parameterization and micrometeorological measurement of air-sea gas transfer. Bound-Lay Meteorol. (2000).

[81] Hare, J. E, Fairall, C. W, Mcgillis, W. R, Edson, J. B, Ward, B, & Wanninkhof, R. Evaluation of the National Oceanic and Atmospheric Administration/Coupled-Ocean Atmospheric Response Experiment (NOAA/COARE) air-sea gas transfer parameterization using GasEx data. J Geophys Res-Oceans. (2004). C8).

[82] Frew, N. M, Bock, E. J, Schimpf, U, Hara, T, Haussecker, H, Edson, J. B, et al. Air-sea gas transfer: Its dependence on wind stress, small-scale roughness, and surface films. J Geophys Res-Oceans. (2004). C8).

[83] Frew, N. M, Glover, D. M, Bock, E. J, & Mccue, S. J. A new approach to estimation of global air-sea gas transfer velocity fields using dual-frequency altimeter backscatter. J Geophys Res-Oceans. (2007). C11).

[84] Asher, W. E, Karle, L. M, Higgins, B. J, Farley, P. J, Monahan, E. C, & Leifer, I. S. The influence of bubble plumes on air-seawater gas transfer velocities. J Geophys Res-Oceans. (1996). C5):12027-41.

[85] Kitaigorodskii, S. A. On the field dynamical theory of turbulent gas transfer across an air-sea interface in the presence of breaking waves. J Phys Oceanogr. (1984). , 14, 960-72.

[86] Farmer, D. M, Mcneil, C. L, & Johnson, B. D. Evidence for the Importance of Bubbles in Increasing Air Sea Gas Flux. Nature. (1993). , 361(6413), 620-3.

[87] Zhang, W. Q, Perrie, W, & Vagle, S. Impacts of winter storms on air-sea gas exchange. Geophys Res Lett. (2006).

[88] Mcneil, C, & Asaro, D. E. Parameterization of air-sea gas fluxes at extreme wind speeds. J Marine Syst. (2007).

[89] Woolf, D. K, & Thorpe, S. A. Bubbles and the Air-Sea Exchange of Gases in near-Saturation Conditions. J Mar Res. (1991). , 49(3), 435-66.

[90] Woolf, D. K. Bubbles and the Air-Sea Transfer Velocity of Gases. Atmos Ocean. (1993). , 31(4), 517-40.

[91] Woolf, D. K. Parametrization of gas transfer velocities and sea-state-dependent wave breaking. Tellus B. (2005). , 57(2), 87-94.

[92] Woolf, D. K. Bubbles and their role in gas exchange. In: Liss PS, Duce RA, editors. The sea surface and global changes: Cambridge University Press, Cambridge; (1997). , 173-205.

[93] Woolf, D. K, Leifer, I. S, Nightingale, P. D, Rhee, T. S, Bowyer, P, Caulliez, G, et al. Modelling of bubble-mediated gas transfer: Fundamental principles and a laboratory test. J Marine Syst. (2007).

[94] Wanninkhof, R, Asher, W, & Monahan, E. C. The influence of bubbles on air-water gas exchange results from gas transfer experiments during WABEX-93. In: Jähne B, Monahan EC, editors. Air-water gas transfer: AEON Verlag & Studio, Hanau; (1995). , 239-254.

[95] Zhao, D, Toba, Y, Suzuki, Y, & Komori, S. Effect of wind waves on air-sea gas exchange: proposal of an overall CO2 transfer velocity formula as a function of breaking-wave parameter. Tellus B. (2003). , 55(2), 478-87.

[96] Broecker, H. C, Petermann, J, & Siems, W. The influence of wind on CO_2 exchange in a wind wave tunnel, including the effects of monolayers. J Mar Res. (1978). , 36, 595-610.

[97] Asher, W. E. The sea-surface microlayer and its effects on global air-sea gas transfer. In: Liss PS, Duce RA, editors. The sea surface microlayer and global change: Cambridge University Press; (1997). , 251-286.

[98] Tsai, W. T. Impact of a surfactant on a turbulent shear layer under the air-sea interface. J Geophys Res-Oceans. (1996). C12):28557-68.

[99] Tsai, W. T. Effects of surfactant on free-surface turbulent shear flow. Int Commun Heat Mass. (1996). , 23(8), 1087-95.

[100] Tsai, W. T. Yue DKP. Effects of Soluble and Insoluble Surfactant on Laminar Interactions of Vortical Flows with a Free-Surface. J Fluid Mech. (1995). , 289, 315-49.

[101] Tsai, W. T, & Liu, K. K. An assessment of the effect of sea surface surfactant on global atmosphere-ocean CO2 flux. J Geophys Res-Oceans. (2003). C4).

[102] Saylor, J. R, Smith, G. B, & Flack, K. A. The effect of a surfactant monolayer on the temperature field of a water surface undergoing evaporation. Int J Heat Mass Tran. (2000). , 43(17), 3073-86.

[103] Goldman, J. C, Dennett, M. R, & Frew, N. M. Surfactant Effects on Air Sea Gas-Exchange under Turbulent Conditions. Deep-Sea Res. (1988). , 35(12), 1953-70.

[104] Frew, N. M, Goldman, J. C, Dennett, M. R, & Johnson, A. S. Impact of Phytoplankton-Generated Surfactants on Air-Sea Gas-Exchange. J Geophys Res-Oceans. (1990). C3):3337-52.

[105] Bock, E. J, Hara, T, Frew, N. M, & Mcgillis, W. R. Relationship between air-sea gas transfer and short wind waves. J Geophys Res-Oceans. (1999). C11):25821-31.

[106] Ho, D. T, Bliven, L. F, Wanninkhof, R, & Schlosser, P. The effect of rain on air-water gas exchange. Tellus B. (1997). , 49(2), 149-58.

[107] Ho, D. T, Veron, F, Harrison, E, Bliven, L. F, Scott, N, & Mcgillis, W. R. The combined effect of rain and wind on air-water gas exchange: A feasibility study. J Marine Syst. (2007).

[108] Ho, D. T, Zappa, C. J, Mcgillis, W. R, Bliven, L. F, Ward, B, et al. Influence of rain on air-sea gas exchange: Lessons from a model ocean. J Geophys Res-Oceans. (2004). C8).

[109] Zappa, C. J, Ho, D. T, Mcgillis, W. R, & Banner, M. L. Dacey JWH, Bliven LF, et al. Rain-induced turbulence and air-sea gas transfer. J Geophys Res-Oceans. (2009).

[110] Ho, D. T, Asher, W. E, Bliven, L. F, Schlosser, P, & Gordan, E. L. On mechanisms of rain-induced air-water gas exchange. J Geophys Res-Oceans. (2000). C10):24045-57.

[111] Takagaki, N, & Komori, S. Effects of rainfall on mass transfer across the air-water interface. J Geophys Res-Oceans. (2007). C6).

[112] Schlussel, P, Soloviev, A. V, & Emery, W. J. Cool and freshwater skin of the ocean during rainfall. Bound-Lay Meteorol. (1997). , 82(3), 437-72.

[113] Lorke, A, & Peeters, F. Toward a unified scaling relation for interfacial fluxes. J Phys Oceanogr. (2006). , 36(5), 955-61.

[114] Wang, B, Liao, Q, Xiao, J, & Bootsma, H. A. A free-floating PIV system: measurements of small-scale turbulence under the wind wave surface. J Atmos Ocean Tech. (2013). in print

[115] Monahan, A. H. The probability distribution of sea surface wind speeds. Part 1: Theory and SeaWinds observations. J Climate. (2006). , 19(4), 497-520.

Novel Tools for the Evaluation of the Health Status of Coral Reefs Ecosystems and for the Prediction of Their Biodiversity in the Face of Climatic Changes

Stéphane La Barre

Additional information is available at the end of the chapter

1. Introduction

Coral reefs concentrate between one quarter and one third of the total marine biodiversity, according to different estimates, though they only cover about 0.1% of the global oceanic surface and are confined to tropical and sub-tropical latitudes. The extraordinary diversity in invertebrates and fish species is often compared to that of arthropods and vertebrates of tropical primary rainforests. Half a billion humans depend partly or totally on the goods and services provided by coral reef ecosystems.

However, coral reefs are now recognized as being among the most fragile of all environments in the face of localized anthropic pressures (overfishing and various atmospheric and water pollutions) and of their climatic consequences of planetary dimensions. Within the last 30 years, about 20% of all coral reefs have being totally destroyed while another estimated 60% are damaged to some extent – a few beyond recovery - and only 20% can still be regarded as unharmed.

With a human population growing at faster rates in developing nations, biodiversity concerns are conflicting with the pressures to exploit local mineral resources and to develop agricultural and seafood production. While residents often deplore the gradual changes in the natural habitats in which they were brought up, socio-economic pressures originally due to post-colonial emancipation and now to global trade tend to resist any policy that may be considered restrictive or receding. Almost every aspect of this economic growth relies on fossil fuel energy consumption and to a lesser extent on vegetal biomass burning, which links development with climatic changes, and also with the generation of polluting wastes.

Educational programs and non-government associations propose alternatives to established practices in habitat and resource management and waste disposal by end users. Research scientists continue to explore natural biodiversity in remote pristine environments (especially *biodiversity hotspots*, a very useful conservation-promoting concept) and to observe its losses in degrading habitats. Programs on bioremediation of impacted sites are attracting funds, and classification of natural habitats as protected sites is gaining public support. Yet the scientific community has very little leverage on the decision-making of potentially impacting industrial, commercial and urban development projects, and on the say-so broadcast by their promoters in the media. In-house consultancy is often hampered by the lack of appropriate analytical tools, and reports are often biased in favor of employment and cash return prospects.

In a previous chapter of this series on biodiversity [1] the various ecological consequences of climatic changes, of chemical and microbial pollutions and of overexploitation of natural resources have been reviewed for coral reef ecosystems. Suggestions have been made on the basis of recent publications by experts on various subjects, as to how modern techniques and innovative approaches could be used appropriately to complement the above initiatives.

In this chapter, a holistic concept is proposed that (i) integrates cutting-edge molecular research and standard technologies with field sampling and laboratory simulations of natural habitats (ii) using holobiont-based *sentinel* systems, (iii) into a single tool that "shows evidence" of ongoing degradation rather than aftermath "score loss". Corrective action can then be taken in specific directions before no-return limits have been reached and total ecosystem collapse is on the way.

2. Biodiversity is our responsibility: The future of mankind depends on it

2.1. If the origins of life remain controversial, biodiversity is a miracle of sorts

Basic forms of life like bacteria are reputedly capable of withstanding extreme conditions, to the point that scientists of repute such as Sir Francis Crick, Carl Sagan or Stephen Hawking have speculated on an alien origin to terrestrial life, which is now held as a tenet of the modern panspermia theory by some exobiologists. Dormant bacterial spores or alike would have been seeded on our planet, possibly from different sources and at different times. Those that encountered favorable "starting" conditions, supposedly in the chemolithotrophic environments around oceanic ridges, would have initiated the evolutionary scenario we know. Posed under such terms, the true *origin of life* escapes our observation, but the *origin of biodiversity* does not, as it remains intrinsically terrestrial.

Satellite views of the ionized portion of our atmosphere show it as a barely perceptible glow that outlines the shape of our planet against the black outer space background. Just under it, blue expanses of oceanic waters spread as a delicate film less than one-thousand times thinner than the supporting "blue marble", but over two thirds of its surface (Figure1). Life forms occur at the sea-air and soil-air interfaces, just where geoclimatic fluxes and exchanges are the most rapid and subjected to biotic influences. Thus marine biodiversity thrives under specific

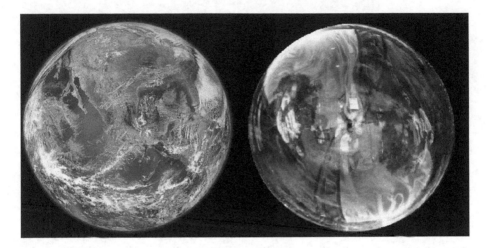

Figure 1. (left) - Composite image of Planet Earth viewed from orbiting satellite Suomi NPP on January 4, 2012 (http://npp.gsfc.nasa.gov/), scaled here at approximately 5×10^{-9}; (right) -The biosphere is about one-thousandth the diameter of the planet – just as thin as the surface of a soap bubble that size, a metaphor illustrating its instability and fragility.

conditions that are only found between the oxygen-rich lower atmosphere (at cloud canopy level on Figure 1), and the sunlit oceanic surfaces.

One-third of all known marine species is concentrated on confettis occupying one-thousandth of the oceanic surfaces, also thanks to favorable conditions afforded by land influence, i.e. marine volcanoes and continental shelves under tropical-subtropical latitudes. This circumstantial miracle is called coral reefs.

2.2. Extinction events are natural over long periods after which biodiversity has to reinvent itself

Any lasting change in the biogeochemistry of any of the three components (atmosphere, seawater and land) will disrupt the interfacial equilibrium that supports the many thousands of life forms that interact constantly within an ecosystem. Mass species extinctions occurred several times (6 or 7 described) in the history of our planet since it became life-supporting, every time followed with new and better adapted life forms and a biodiversity climax attained after long periods of environmental stability. Changes in soil mineral strata indicate the occurrence of biodiversity-modifying events such as occupation by seas or the occurrence of an ice-age. Discrete organic layers may indicate the presence of a tropical rainforest or of a dry land savannah. Datable fossil evidence within these strata, together with paleontological reconstructions, point out the floristic and faunistic peculiarities of the times. Core drills in ice provide datable evidence of biogeoclimatic episodes within the last few millennia, while core drills in massive scleractinian corals give accurate calendar-like records of recurrent or of accidental climatic events affecting their biotope.

Speciation usually goes along with occupation of new territories and new habitats, the first colonizers having acquired the necessary adaptations to cope with evolving external demands – the Cambrian explosion (545 million years ago) being the most dramatic example of such adaptive diversification at all scales.

Along with this, evidences of accidental episodes of massive species extinctions are witnessed by the sudden "disappearance" of terrestrial and of marine life, that are attributable to tectonic, telluric or meteoritic impacts and to their profound and lasting climatologic and geochemical consequences. The most significant mass extinction is undoubtedly the Permian-Triassic Great Dying [2] where a 96% loss of all non-microbial marine life occurred within ten million years. The precise causes of mass extinction events may be in connection with continuous tectonic movements with their telluric and volcanic outbreaks and their climatic consequences, to collisions with meteoritic bodies, and to a lesser extent to the appearance of dominant predators, parasites or microbial diseases, or to combinations thereof. Common to many extinction events, however, is the massive release of greenhouse and of toxic gases (carbon dioxide, methane, hydrogen sulfide etc.). The water solubility of CO_2 being nearly 30 times that of oxygen, water acidification occurs that impacts preferentially all calcifying organisms with low metabolic rates and weak respiratory systems: most coral genera died out during the Great Dying, along with calcareous sponges, calcifying algae, echinoderms, bryozoans etc. [3].

Interestingly, profound taxonomic changes in all major phyla seem to follow extinction events, resulting in a better adapted biodiversity. Nothing is known, however, on the consequence of such changes in microbial life or on the putative role microbial associates had on the *reinvention* (understand: adaptive evolution) of new species. As Falkowski [4] puts it, animals and plants are merely new incarnations of ancient metabolic processes, but the ultimate key to biodiversity may be held by bacteria ferrying the set of core genes that are necessary for life to express itself.

2.3. Brutal human influence may lead to hazardous extinction events

Man is of very recent occurrence in the history of terrestrial life, and until the advent of the industrial age (especially since the middle of the 19th century) his planetary influence was minimal, and the destructive potential of its inventions was purely local. Since then, his population has increased seven fold while consuming or exploiting 42% of the terrestrial net primary production [5]. The single most environment-impacting activity today is the production of energy from fossil fuel, and the single most mechanically destructive invention is that of nuclear weapons. More insidious are the thousands of new chemical species (mostly organic) that are produced and improperly disposed of once used, and the new strains of mutant microorganisms that pop up unpredictably and threaten to create worldwide pandemias affecting both wildlife and humans.

Without reviewing the subject of negative traces of human doings (see [1]), a reflection on the accelerating pace of carbon volatilization in the atmosphere is necessary to appreciate its effect on marine biota and on coral reef ecosystems in particular. In one year, we extract the equivalent of one million years' worth of fossil fuel natural production [4]. In other words, using primitive technology e.g. internal combustion engines, we enrich our atmosphere in carbon dioxide at a pace at which terrestrial and marine autotrophs will not be able to keep up with in their efforts to create new biomass and quench free carbon excess and reduce seawater acidification rates.

3. Coral reef management: between the economy's hammer and the climate's anvil

Ecosystems at immediate risk are biodiversity hotspots (primary rainforests and land-connected reefs systems), but most ecosystems with endemic species at all latitudes will be profoundly affected by acclimation of competing aliens. Being able to analyze, quantify and predict changes is the first step to avoid losing control... indeed environmental scientists should endorse the role of a general practice physician making an overall check-up on an ailing patient and prescribing a course of medications and exercises for recovery, or that of an investigator using state-of-the-art forensics such as DNA amplification to evidence causes of a crime and provide evidence to lawyers. This will become a necessity, beyond economic pressures worldwide (q.v. [6]).

3.1. Climate changes have profound biogeoclimatic consequences

Climatic changes act globally, and the effects of naturally occurring destructive episodes are now superseded by those due to ever-growing man-induced emissions of various sorts. The latter include carbon, nitrogen or sulfur transfers that are generated essentially through the combustion of fossil organic matter or of agricultural or livestock breeding activities, and manifesting themselves differently according to their biogeochemical state.

i.   First, their volatilization as simple molecular species which are responsible for temperature rises through the glasshouse shielding effect, while halocarbon emissions tend to destroy the protective ozone layer of the upper atmosphere. Elevated surface water temperatures and genotoxic radiations are responsible for the bleaching effect and cellular stresses observed on subtidal corals, and death may follow if the exposures are lasting. Also, strong seasonal evaporation of large volumes of surface waters leads to atmospheric pressure instabilities, resulting in the occurrence of more frequent and more severe hurricanes. The mechanical damage due to wave action and to the sudden input of large volumes of freshwater and alien material (silt and debris) in coastal zones may entirely destroy some reef portions in single episodes;

ii.   Second, the restitution of most of the above molecules at the air-water interface and their incidence on biomineralization of many marine invertebrates and of their larvae, and of coralline algae; In the 22nd century, seawater acidification may be as large as 0.5 pH units below the pH levels recorded in early 20th century. The extent of seawater acidification due to a shift in bioavailability of HCO_3^- concentration, if still debated in 2012, may shortly become dramatic for the survival of calcifying invertebrates (corals, mollusks, echinoderms, some sponges etc. and their larvae) and of coralline algae.

3.2. Societal changes: think globally, act locally

Socio-economic determinants are linked to the development of industries, commerce and urbanism, their primary effects being direct, localized and traceable.

- industrial activities typically generate important volumes of polluting agents that dissipate during the processing stages. Furthermore, residues are managed according to criteria of productivity and cost-effectiveness i.e. with little concern about recycling alternatives, and about speciation into toxicants that make their way from the discharge areas into waterways and eventually into the sea. Ore excavation or grading of metal-rich top soil cause faunistic and floristic deteriorations, i.e. loss of endemic wildlife, facilitate bioerosion, cause the constant remodeling of river mouths with the finer particles and destroy almost life forms with the accumulation of heavier deposits in basins. In the sea, bioaccumulation of organic pollutants by primary consumers can expectedly reach very high concentrations in top predators, e.g. lipid-bound chlorinated biphenyls. Heavy metals that leach out from anti fouling formulations are accumulated by filter-feeding invertebrates, and take a heavy toll on the reproductive success of many life forms found in intertidal and subtidal life forms around harbor waters;

- intensive agriculture and farming generate various type of pollution, (i) enrichment by fertilizers and farm animals disjecta, (ii) pesticides and systemic herbicides, (iii) veterinary products, hormones that alter the quality of underground water reserves, while excess runoff benefits undesirable primary production in coastal zones, e.g. "green tides" and microalgal blooms;

- global commerce favors dissemination of alien species by shifting large amounts of ballast water by container ships and cleaning of hulls;

- urban development in tropical zones eliminates mangrove forests and seagrass beds that are important coastal component of fringing reefs. Shifting into the sea of very large volumes of untreated sewerage generated on land causes eutrophication in coastal reef systems with the introduction of novel microbial diseases.

3.3. Scientists should play a central role in biodiversity issues

At present, there are no reliable estimates of the rate at which coral reef biodiversity is going to be affected by climate and anthropic forcings by year 2100. The various proxies used by climatologists and reef ecologists predict two types of scenarios: (i) short term collapse of tropical reef biodiversity as early as 2050 if current trends are not curbed, (ii) a good level of resilience if present-day conditions are maintained as an absolute maximum, allowing photosymbiotic systems to adapt gradually. By limiting direct interference (q.v. *overfishing, algal proliferation, alien species*) that introduce phase-shifts in the complex and otherwise self-regulated food chains, acclimatization of most common coral species can be made possible in the face of climatic changes. The second alternative represents the only and narrow line of action, knowing that it will not be realistic to completely stabilize CO_2 emissions to safe levels, i.e. natural quenching capacity of excess by phytoplankton.

In 2012, scientific expertise is mostly directed to the discovery and documentation of new species in remote or in hitherto neglected regions, or to report the loss of biodiversity in areas impacted by human activities. Consulting is sought for the delineation of protected areas and establishing quotas for fishing. Science-based management is necessary, in particular in regions suffering from direct and rapidly growing human influence [7]. Particularly at risk are young developing nations and small islands that are placed under tremendous pressure from major industrial countries to exploit and export their natural resources – a socio-political scenario that tends to divide local populations and sometimes accelerate environmental issues. Thus, care must be taken of the "growth crises" of these culturally fragile ethnic groups undergoing these economic pressures, and a premium should be placed on the management of their natural patrimony, in which their culture is firmly rooted.

What comes out most often in recent discussion groups is: (i) that new monitoring tools, new environmental technologies and new models must be developed within a consensus mode between interested parties (ii) scientific expertise and field experience must be federated into autonomous consortia, (iii) implementation stages must follow a rigorously coordinated and step-wise approach. The present feeling is that scientific expertise is only marginally called upon and often misquoted in the decision-making of matters of economic development and urbanization in developing countries. By following the above recommendations, habilitation of the scientist as an advisor or as a mediator playing a crucial role in decision-making will be greatly facilitating the social dialogue on ecological matters.

4. State-of-the-art: New concepts, novel approaches, cutting-edge technology etc...

How to find a suitable management compromise between climate and human forcings without altering the economy and the social development of developing nations? This is everybody's concern: entrepreneurs, merchants, economists, businessmen, politicians, educators, end-users and consumers and of course, scientists.

From the scientist's point of view, each identifiable environmental issue has its specific set of solutions, and should be treated with the same care as an outpatient's condition is in a hospital. The physician makes a global assessment and recommends that a series of analyses be made by specialists before he confirms his diagnosis and prescribes an adapted treatment. In most cases, the patient will recover successfully – sometimes he has to be hospitalized for some time, and on rare occasions he may not leave the hospital alive.

After an overall check-up in which the probable stressor(s) is (are) identified, the environmental investigator will hand the case over to specialists who will each carry out a set of biochemical, molecular, microbiological and imaging tests on specific model organisms, on their tissues, cells, body fluids and associated microbes. Once the diagnosis is confirmed, a preliminary report is made that may contain special recommendations, followed by regular visits to evaluate the resilience of the system and the potential for the recovery of its lost biodiversity.

To be able to adapt this medical approach to an ailing ecosystem, the environmentalist needs to find representative biological study models (i.e. sentinel species that are sensitive to the stressors, but not to the point of immediate eradication at the onset of a mild exposure). He also needs to study characteristic and observable symptoms, their evolution and their succession. Early markers of an organism exposed to a stressor can be detected using functional genomics with suitable molecular tools, and the evolution of the responses can be followed thenceforth. For instance, an organism undergoing abiotic stress is usually more prone to microbial infection than a conspecific control organism, calling for physiological and bacteriological / fungal / viral / parasitic analyses. Unfortunately, very few field investigators are in a position to use these tools routinely, let alone in association - especially molecular tools that have been adapted from the medical world to specific biological models, and only for research.

4.1. Shifting from a consumer-minded to a conservation-minded attitude

Coral reef organisms tend to live in close association in order to gain optimal access to essential resources such as hard substratum, access to light, appropriate food etc., the co-occurrence of which is limited in shallow tropical waters [8], which are naturally efficient at nutrient cycling (oligotrophic regime). These constraints are reflected in the highly sophisticated communal assemblages formed by coralline and fleshy algae, invertebrates, vertebrates, protists etc. which communicate essentially via surface-to-surface or distant interactions of immunological and/or chemical nature. All forms of associations are encountered, ranging from obligate parasitism to symbiosis via commensalism, from prey hunting to filter feeding via surface browsing, etc.

The scientific literature on coral reef biology has traditionally emphasized on competitive aspects between members sharing a same habitat, especially in connection with secondary metabolites emitted by sessile or sedentary invertebrates and having allelopathic or growth inhibiting activities [8,9]. Only recently have the cooperative and functional aspects of interspecies and cross - phyletic communication been explored - stepping from a more "medical" attitude (i.e. pharmacologically-oriented) to a more "ecological" one (i.e. conservation-oriented). Discoveries such as bacterial inter - communication via quorum-sensing signals, biofilm studies, and of course the progresses made in genomics have greatly contributed to this attitude change (e.g. the coral probiotic hypothesis of Reshef et al. [10].

In the eighties, the term *biodiversity* was publicized by [11] to account for the need to evaluate and to manage biological resources in endangered habitats. It gained immediate recognition as a key concept in life sciences, and public adoption followed relayed by media channels, e.g. "citizen science" in which data collection by non-scientists was encouraged to feed statistics on endangered species, as well as public initiation to scientific issues [12]. Ecosystem services and human well-being oriented approaches in general, while often assimilated to biodiversity-conservation strategies, may not ultimately pursue the same goals, calling for trade-offs in difficult issues [13].

The biodiversity concept certainly helped in the funding of sampling expeditions in diversity *hotspots*, another term useful to conservationists and to taxonomists. The Tree of Life (http://

tolweb.org/tree/) is a collaborative web project in which over 10,000 web pages provide information about extant groups of organisms and their evolutionary history.

Rapid progress in molecular science, and genomics in particular, has made it theoretically possible to retrieve useful information from crude environmental samples, i.e. doing away with cultivation restrictions and going beyond conventional taxonomy based on morphological traits. However, metagenomic characterization of marine microorganisms from the plankton, from biofilms or associated with macrobiota and their exsudates, generates phenomenal volumes of data from which crucial information extraction is a difficult and costly task for biocomputing specialists. Advocated by some as a new paradigm for biodiversity studies [14], "data-intensive science" taking a naïve (data-driven, i.e. non-theory-based) approach looks for truly novel and surprising patterns that are "born from the data". But because of meta-data sorting problems and costs, and since environmental problems cannot wait for new hypotheses to emerge, others think that "knowledge-based science" can at least confront existing hypotheses against the meta-data background and guide investigators into detecting novel information [15].

4.2. The holobiont as an evolutionary concept, the *extended holobiont* as a conservation concept

Closer to hands-on science, genomists have coined a very useful concept, originaly to account for the functional dynamics associated with bioconstructing organisms such as coral, sponges, or coralline algae in tropical marine ecosystems: that of the *holobiont*, an entity that includes the host (basibiont) and its microbial associates (microbionts) – the *hologenome* being the sum total of the associated genomes [16]. Photosymbiotic systems may function of both autotrophic and heterotrophic modes with the assistance of, respectively, symbiotic microalgae or cyanobacteria during the day, and element-recycling bacteria inside bacteriocytes or associated with the mucus, tissues or skeletal cavities followed by directed uptake at night. This dual mode of operation is conferring an adaptive advantage to short-term disturbances [17].

The hologenome theory was later generalized to terrestrial eukaryotic-prokaryotic systems, including man and its microflora [18], in order to account for the coevolutionay and cross-kingdom aspects of symbiosis [19].

Finally, the holobiont is also a practical concept insofar as it can be conveniently transposed from its natural sites to a microcosm or mesocosm aquarium setup, in order to evaluate stress impacts on both the host and its associated microbiota.

The concept of *extended holobiont* is proposed here, as the urge is now felt to develop practical tools to directly address biodiversity issues. In their definition of the coral holobiont, [20] include cryptic associates that are structurally associated with the host (fungi, encrusting sponges and algae, protists). In order to maintain some evolutionary coherence between the holobiont *sensu stricto* and the hologenome concepts of Rosenberg, we may include the dozens of specialized and mobile life forms that live in trophic or parasitic or commensal relationship with the host (e.g. little crabs, shrimps, echinoderms, ascidians, polychaete worms, shelled or naked mollusks, planarians, fish, crustose algae, fungi, foraminifera, etc.) into a wider concept

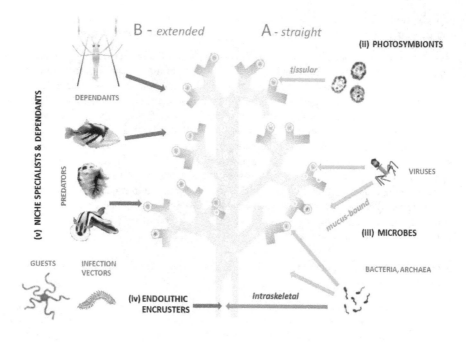

Figure 2. (right) A - Representation of the coral holobiont *sensu stricto* i.e. basically a functional 3-way photosymbiotic system with (i) the sedentary or sessile basibiont (center) and its mucosphere (pale grey halo) with (ii) its photosymbiotic associates and (iii) prokaryotic microbiome; (left) B - The *extended holobiont*, which may include (iv) endolithic associates (fungi, encrusting sponges and algae, protists) and (v) biodiversity that lives in association with the coral and may be critically affected by the loss of the host

which may account for the host- dependant biodiversity. We propose to refer to it as the *extended holobiont*.

4.3. The "omics" or high-dimensional biology revolution

"Omics" technologies are primarily aimed at the universal detection of genes (*genomics*), mRNA (*transcriptomics*), proteins (*proteomics*) and metabolites (*metabolomics*) in a specific biological sample [21]. Omics can be used in a wide variety of applications, ranging from biomedical to environmental, from biotechnological to ethical, from single-cell to ecosystem-wide, from systems biology to phylogeny. In short, "omics" technologies adopt a holistic view of the molecules that make up a cell, tissue or organism. This integrative approach, together with the "extended holobiont" as a functionally comprehensive biological model, provides a useful conceptual framework when dealing with complex environmental issues.

Transcriptomics - Perhaps the single most informative approach to studying the fitness of a test organism being subjected to an environmental stress is transcriptomics. In essence, the aim is

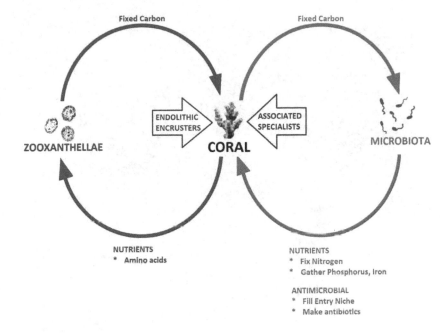

Figure 3. Relationships between the components of the coral holobiont system (developed from [20], with the addition of cryptic organisms and of associates (the *extended holobiont*)

(i) to catalogue all species of gene transcripts, including mRNAs, non-coding RNAs and small
RNAs (non-coding bacterial RNAs), (ii) to determine the transcriptional structure of genes in
terms of their start sites, 5′ or 3′ ends, splicing patterns and other post-transcriptional modifi-
cations, and (iii) to quantify the changing expression levels of each transcript during development and under different conditions [22]. The set of coding transcripts from activated genes
will encode proteins as products of the ribosomal assembly line. Initially, complementary DNA
(cDNA) copies are created from the mRNA templates, from which two analytical techniques
have been developed: (i) hybridization-based and (ii) sequence-based. In hybridization
transcriptomics, DNA microarrays composed of collections of a collection of microscopic DNA
probes are used to hybridize cDNA targets - here a gene is activated when a spot is highlighted
with a visual label, allowing quantification as well as characterization. Yet, the method relies
on existing knowledge about the genome sequence. Modern sequence-based methods (RNA-
seq) determine cDNA sequences directly, the latter (30-400 bp.) being attached terminally to
adaptors. Each short sequence is aligned with the reference genome or transcriptome and
classifies as exon read, junction reads and poly(A) end reads, from which a base-resolution
expression profile is generated. A critical comparison between the different transcriptomic
techniques is proposed in [22], with an emphasis on RNA-seq.

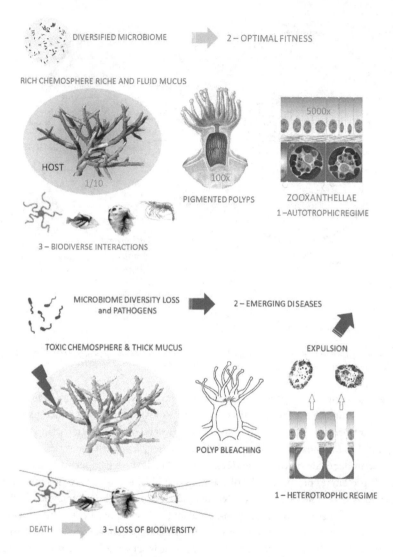

Figure 4. The extended holobiont of a scleractinian sentinel system under contrasting conditions: (1) top: under optimal conditions, the fit colony produces a fluid mucus which hosts a well-diversified and functional microbial flora; the polyps are colourful thanks to a dense population of symbiotic zooxanthellae that colonize the endodermal layer and provide the substantial autrotrophic part of the energy budget of the holobiont. Eukaryotic associates are numerous, engaging into a network of interactions centered on the host. (2) Bottom: under lasting stress, e.g. sea water temperature above 32°C, the colony may produce a thicker mucus, in which a growing population of pathogenic bacteria tends to displace less competitive strains. Zooxanthellae leave the polyp tissues as a result of loss of immune recognition, and no longer provide photosynthates to the holobiont system which survives on heterotrophic/direct diffusion regime. Rate of associate biodiversity loss depends on the type of relationship with the host, from obligate to occasional

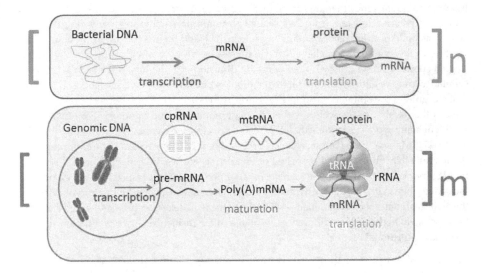

Figure 5. The total transcriptome of a holobiont system (1) top: the total set of bacterial transcripts from a population of *n* cells (2)bottom: the total set of eukaryotic transcripts from a population of *m* cells. Separating the coding RNAs from the non coding ones is made on the basis of the presence or absence of a poly-adenosine monophosphate tail, but only the final mRNAs are usually selected in transcriptome analyses

Stress reactions have been studied on of components of the coral holobiont. In [23] De Salvo and collaborators performed a medium-scale micro transcriptomics experiment on the Carribean coral *Montastrea flaveolata* undergoing thermal stress and bleaching vs. unexposed conspecifics. Their complementary microarray experiment containing 1310 genes revealed that thermal stress and bleaching in this species affected the following processes: oxidative stress, Ca^{2+} homeostasis, cytoskeletal organization, cell death, calcification, metabolisms, protein synthesis, heat-shock protein activity and transposon activity. A dedicated transcriptome database was recently produced on *Pocillopora damicornis*, a ubiquitous and environmentally sensitive scleractinian [24], [http://cnidarians.bu.edu/PocilloporaBase]

Proteomics - Physical or chemical stress (i) inactivates or down-regulates many genes, including many housekeeping genes, (ii) while up-regulating stress genes that perform orchestrated induction of key proteins necessary for cellular protein repair. Stress expressions of organisms, or comparative analysis of ecotypes of a species naturally found in contrasting environments, can highlight the production of stress-related enzymes, such as heat-shock proteins. Heat stress does denature proteins, causing weakening of polar bonds, unfolding and exposure of core hydrophobic groups. Beyond the cell's tolerance, heat stress will cause its death. Thus the cellular stress response (heat-shock response) protects organisms from damage resulting from critical fluctuations of e.g. heat, UV light, trace metals, and xenobiotics. Stress genes are activated to rapidly synthesize stress proteins, a universal and highly conserved response (from bacteria to humans in which similar roles are played).

Indeed, candidate genes that are directly implicated (e.g. corals' response to stress) or functionally interconnected (e.g. to genes related to immunity) with cnidarians-dinoflagellate symbiosis point out the major role of key proteins [25], and their fast-evolving adaptation in the face of environmental challenges. Oxidative stress in zooxanthellae produces reactive oxygen species (ROS) with hydrogen peroxide diffusion into the host cell which activates a cellular cascade resulting into the photosymbiont expulsion and polyp bleaching. It may be that recognition of a suitable zooxanthellae clade by the coral host is a selective process during which other strains are actively expelled through immunity and apoptosis [26], the photosymbiont being more susceptible than the host to e.g. elevated temperatures and possibly UV levels [27]. Massive and laminar species are supposedly more resistant to environmental fluctuations than shorter-lived branched or encrusting species. Investigators in [28] showed that laminar corals (using *Turbinaria reniformis*) undergo transient Hsp60 heat-shock protein induction under either light or thermal overexposure, and prolonged induction of Hsp60 if the two treatments were combined – whereas branched *Stylophora pistillata* were relatively unresponsive, highlitghting differences in potential for resilience between different growth forms and nutritional regime.

Metabolomics - Notwithstanding their dedicated specialist predators and parasites, scleractinian corals in general are poor producers of secondary (i.e. defence) metabolites since they enjoy protection from a biomineralized skeleton into which polyps can retract – a situation quite different to that of soft corals (alcyonarians) which produce a whole range of cyclized terpenoids against predators, for space occupation and larval protection. However, the presence of several classes of compounds, naturally produced or stress-elicited, has been described for massive coral species (e.g. *Porites* and *Montipora*), and in a number of branched forms. In the former, thermal stress followed by pathogen-associated molecular patterns (PAMPS) triggered defensive responses againt alien threat as well as protective responses, e.g. phenoloxidase activity (production of melanin pigment [29], antibacterial activity, peroxidase and ROS scavenging (oxidative stress) and fluorescent protein [30]. Branched forms are regarded as more sensitive to thermal stress than massive forms, and produce natural antibiotics. Ectodermal cell of *Pocillopora damicornis* release damicornin, a 40-(residue antimicrobial peptide) in response to non-pathogenic challenges, but its expression is repressed by pathogen *V. corallilyticus* [31]. Various *Pseudoalteromonas* strains are known to produced antibiotics, e.g. diketopiperazines, that help them control the bacterial profile sharing the same host-associated resources [32], especially during changes in bacteriome profiling [33]. Mucus-associated *Pseudoalteromonas* spp. are generally considered efficient at protecting the coral holobiont's defense against potential Gram-positive pathogens [34], and are already known to participate in the antibiotic defense of green algae, against surface fouling organisms [35] and fungal epiphytes [36]. Each of these molecules is a potential chemomarker of stress.

Genomics - Marine eukaryotes span 35 phyla, 14 of which are exclusively marine. Genomics is useful in complementing taxonomy beyond classical morphological keys, in exploring evolutionary traits and similarities of small are large taxa, and in investigating the therapeutic potential of target taxa. Furthermore, molecular barcoding allows accurate assessments of environmental samples (especially protists an prokaryotes) – a very useful tool in estimating biodiversity. Corals genomes include 23677 genes (fewer than human genome with 36036 but also than paramecium genome with 39581 genes). The genomes of several common species of scleractinian corals have been characterized at least partially. *Acropora millepora* [37], *Acropora digitifera* [38, 39], genomes have revealed a more complex than hitherto suspected complexity, especially regarding calcification and innate immune repertoire, though scleractinians seems to have lost the ability to carry out *de novo* synthesis of photoprotective mycosporin-like amino acids (MAA) family of compounds. Innate immune responses provide valuable information as predictors of thermal stress susceptibility and disease. A recent paper by [40] usefully reviews key features in coral immunology (recognition, signalling pathways and effector responses) in the general context of invertebrate immunology. Gene expression markers have been recently developed to monitor early responses to acute heat-light stress in *Porites* species, on the basis of differential expression of Hsp16 and actin genes [41]. Genes involved in the immune response against bacterial pathogens present clear differences in their expression patterns between *Vibrio corallilyticus* – exposed *Pocillopora damicornis* vs. unexposed control conspecifics [42]. Such predictors of stress responses can prove very useful if integrated in monitoring tools.

A comprehensive expressed sequence tags (EST) transcriptomics dataset on the symbiotic zooxanthellae has recently indicated some unique regulatory characteristics not found in free dinoflagellates – once completed and annotated, the complete *Symbiodinium* genome will represent a considerable asset in interpreting the coordinated responses of the coral holobiont under stress [43], beyond completely elucidating the molecular bases of the host-photosymbiont association. A recent worldwide survey spanning 20 years of data on coral-*Symbiodinium* photosystems revealed that the transmission mode correlates positively with photosymbiont specificity, not with coral specificity [44]. The authors found fewer generalist species than specialist species. The former group include the common Pocilloporid scleractinians and several Acroporids that can accommodate more than one zooxanthellae clade, and are more environment-sensitive than e.g. common massive corals which may be more resilient to contemporary stressors, especially those genera that are vertical transmitters such as *Porites* and *Montipora*. These findings point out the necessity to select one sentinel species of each group (branched vs. massive) in order to obtain a less biased assessment of the health status and of the resilience potential of a given reef locality under stress.

The third major functional component of the coral holobiont is the bacterial and archaeal microbiome, also susceptible to large composition shifts during e.g. heat stress [45], as revealed by metagenomic studies. Bacterial consortia are described as host-specific, each profile having its specific and its generalist strains, the relative composition of which being affected by environmental conditions.

Ideally, holobiont-wide analysis will benefit from the combined knowledge of the molecular biology of coral, zooxanthellae and prokaryotes that are necessary to define a fully functional system used as a control. The fitness descriptors will need to take into account (i) geographic variations, e.g. chemotypes for widely dispersed taxa), as well as (ii) short natural environmental fluctuations that fall into the natural physiological tolerance of a given population.

The introduction of a stressor (significant in intensity and/or duration) will allow investigators to model the precise interactions of molecular events that affect the three components of this holobiont, and place the physiological responses of the different parties into a system-wide sequence, from early responses to total collapse. Using a holistic and clinical-like approach [46], issue - specific network models can be created by confronting data collected on "stressed" holobionts against homeostatically regulated "no-stress" conspecific controls.

4.4. Merging "omics" with imaging and physiological / ecotoxicological approaches

Creating a multi-approach and comprehensive tool to evaluate the health status of corals under climatic or direct anthropic threat provides a more robust assessment than when using a single analytical method. The omics revolution is coming of age, and large scale data collection (metadata) are more easily tractable using modern bioinformatics algorithms (next generation sequencing) than before. The number of research papers dealing with molecular aspects of photosymbiosis in corals, with stress responses leading to the host-symbiont rupture, with the detection of early markers of stress (before actual symptoms are physiologically or visually expressed), with the resilience potential of massive vs. branched growth forms, with innate immunity, with the onset and development of bacterial pathogenicity following stress, has increased tremendously within the last few years. Each paper brings its unique and useful light into one of the most fascinating biological phenomenon. As of today, however, scientists are little more than spectators of a dramatic acceleration of the destructive impacts of civilization on the most fragile of all marine ecosystems. Politicians and the media are only implementing or relaying preventive anti-pollution policies, and what high-tech research has to offer to monitor what is actually occurring goes well beyond the understanding of the layman.

What is proposed here is to select the most informative analytical strategy as the core component, molecular biology (omics) being the choice alternative to detect early stress responses, and to complement it with physiological measures that are privileged for evidencing (i) adaptability and (ii) loss of function. Physiological monitoring is important in comparing the tolerance range of corals various growth forms, to short lived or limited stress exposure that do not cause changes in the composition of zooxanthellae [47], and allow for gradual acclimation. This laboratory experiment on *Acropora millepora*, a favourite model, may explain why in some localized coral populations, holobionts were capable of resisting temperatures well above the known tolerance limit for the species. Whatever the analytical method used for physiological measurements on the coral host or on its photosymbionts, the instrumentation must be adapted and the measurements repeatable in order to have strong and scalable metrics both in measuring the level and duration of the stressor, and the responses on behalf of the holobiont component under investigation. Respiration and photosynthesis, oxygen production, calcification rates, mucus rheology, pigmentation etc. can be very useful as complementary analyses along with omics.

Imaging tools have a greater impact to non-scientists, and their can offer excellent visual "proof" of an ongoing stress response, in the field (e.g. time-lapse photography of entire holobiont), or inside the component under examination. For example high-resolution imag mass spectrometry or NanoSIMS can make isotope tracing at single cell level. When linked to molecular visualization methods, such as *in situ* hybridization and antibody labeling, these techniques enable in situ function to be linked to microbial identity and gene expression [48]. Fluorescent pigments, e.g. multi-use pocilloporins, that have a photoprotective and antioxidant roles as well as helping dissipate excess light energy [49] can be detected by classical UV-VIS spectrometry. In fact green fluorescent protein-like (GFP-like) coral fluorescent pigments are routinely used as biomarkers in medicine and can be used to detect coral growth and tissue proliferation, as well as monitoring reactive oxygen species scavenging during stress. Hyperspectral pigment imaging in combination with oxygen profiling provide useful information on competitive interactions between benthic reef organisms, and demonstrate that some turf and fleshy macroalgae can be a constant source of stress for corals, while crustose coralline algae are not [50].

In combination, -omics, physiological/ecotoxicological and imaging tools provide a potentially formidable combination for measuring stress responses in coral holobionts and their separate components.

5. What modern technologies can do for the environment

Ideally, we need a multi approach diagnosis tool focused on the sentinel species undergoing stress, on the profile of its microbial associates, and to be able to estimate the loss of the epibiotic and encrusting macrofauna and flora which lives in association with the living host, i.e. the evolution of the overall biodiversity from the earliest stress symptoms detectable on the host, to its death.

The basic requirements for a diagnostic tool are:

- to consider different components of the holobiont model (i.e. host, photosymbionts and microbiota) and then integrate the results of the different analyses into a single comprehensive "holistic" diagnosis;

- for each analysis and each approach (molecular, microscopic, ecotoxicologic), to be able to define dose/exposure limits of the stressor that correspond to threshold responses along a continuum such as: normal tolerance/acclimation/resilience/no-return/rapid death;

- to propose corrective measures wherever some critical point is reached in one of the above limits.

The basic requirements for a biodiversity-assessment tool are:

- to list the algal and animal species that are usually associated with the sentinel holobiont;

- to categorize each species with respect to its location within the holobiont system (encrusting, epibiotic, mucus-bound, free-living) and to its degree of dependence to the host for each type (predator, commensal, parasite, symbiont).

5.1. Measuring stress responses, diagnosing overall fitness and proposing corrective measures

On the basis of the above requirements for a diagnostic tool, we propose an 8-step procedure to achieve a compromise between experimental robustness and implementation simplicity. The sequence is shown in Fig. 6, and each step is detailed in the text.

Figure 6. The 8-step logical sequence to create a customized tool (the INDICORAL procedure), each step being described in the text

STEP ONE (ISSUE) - Identification of the problem (in the field)

Each biodiversity issue is different, and the first task is to identify the source of the problem. Abiotic stresses should not include first-degree biological interference due to competition, predation or outbreaks of invasive or of alien opportunists, including benthic and pelagic macro and microbiota. On the other hand, we shall consider the departures from "normal" or standard macro and microbiodiversity components resulting from the host species being under abiotic stress, e.g. expulsion or "bail out" of photosymbionts and loss of useful bacterial strains, or disappearance of vegetal and animal associates, which are at the very heart of the biodiversity loss issue. Most common abiotic stressors include: thermal stress, dessication, irradiation, hypo-hypersalinities, silting, heavy metal accumulation, organic compounds, mechanical damage due to wave or wind action and to anchoring. Thus, abiotic stressors can be climatic or pollution bound, chronic or accidental and have their effects combined. It may be necessary to combine field and laboratory (aquarium) studies in which stressors can be analyzed and modulated individually.

STEP TWO (METHODS) - The choice of the adequate holobiont (field)

The choice of a biological taxon may depend on the type of stressor. Corals and coralline algae might be more suitable to evaluate thermal, ultra-violet and acidity stresses, whereas bioac-cumulators such as sponges or bivalve mollusks might be more suitable for silting and heavy metal stresses. Furthermore, a good model is one which is sensitive enough to the stressor, at the same time displaying a range of responses that can be calibrated usefully against different concentrations / exposure times. Finally, the model host-species must be (i) representative of the area under investigation, (ii) common enough for sampling at statistically significant scale and (iii) amenable to aquarium studies.

STEP THREE (METHODS) - How to measure stress in a model holobiont system

Abiotic stress will affect all components of the holobiont diversely and in a network connection manner, with gradual loss of function and accompanying morbidity symptoms. Stress studies will therefore be dealing with the host organism, its photosymbionts, and also with its associated microbiome. This will be achieved through a combination of cutting-edge and classical approaches, e.g. (i) *in aquario*: transcriptomics on the basibiont, microscopy/cytometry and molecular biology on the photosymbiont, (ii) *in the field*: bacterial and viral metagenomic sampling on the associated microbiomes, and ecotoxicology/physiology on the holobiont.

In aquario studies will use a significant number of the selected sentinel species from undisturbed environments, or cloned fragments of same where applicable.

STEP FOUR (DATABASE) - Making a robust set of control data for each analysis

The sentinel holobiont must be sampled in reputedly undisturbed areas, whatever the type of analysis: physiological, metagenomic, taxonomic (associate biodiversity) etc. and natural variations from a statistically significant number of replicates must be recorded.

When dealing with a photosystem holobiont, aquarium studies should consider the three components separately: (i) the holobiont e.g. coral, sponge…, (ii) the photosymbionts, e.g. zooxanthelae, cyanobacteria, (ii) the associated microbes e.g. mucus and tissue-bound bacteria. In each case, a minimum of three (and up to five) independent analyses using different analytical approaches must be undertaken, e.g. one or two –omics, one or two physiological or ecotoxicological, one using imaging etc.

For each analysis, the average values or estimates will set the 0 mark or control score of a future 0-10 scale. This step is crucial and the natural variability must not be too large with scores never exceeding those of responses to mild-severe stresses.

STEP FIVE (DATABASE) – Set experimental calibration scales (Fig. 7)

A useful way to calibrate a response scale *in aquario* is to expose the holobiont or its separate components to different stress regimes (concentration and duration) and establish a lethal value (10 on the 0-10 scale) and in between define threshold values (first symptoms, loss of photosymbionts, loss of microbial biodiversity, irreversible loss of function, death < 12h). Each analytical method will highlight a sequence of events along this "control to immediate death" gradient.

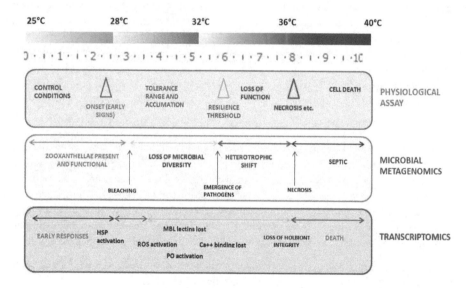

Figure 7. Studying responses of a coral holobiont during experiment heat/light stress using different analytical methods, and establishing a 0-10 scale of responses with critical values. This diagram is indicative only (values and sequences are fictional)

STEP SIX (ANALYSIS) - Compare stress vs. standard profiles

Once each experimental scale is established per analysis, environmental samples can be rated by attributing an average performance score based on n replicates.

STEP SEVEN (TREATMENT) - The single grid of impact or radar chart (Fig. 8 top and middle)

Then scores for each of the 3-5 analyses on each of the biological components will be reported to a single grid of impact, e.g. on a coral (5), its photosymbionts (4) and its microbiome (3).

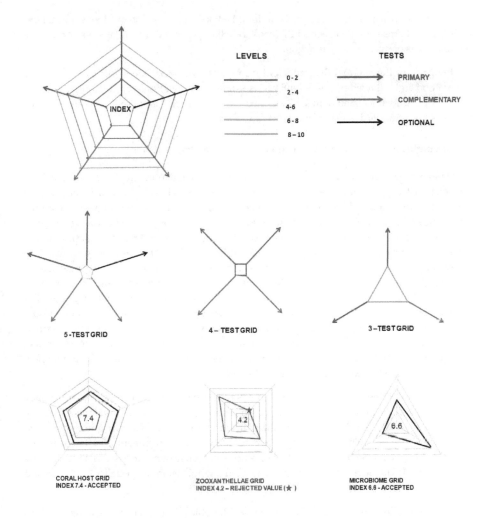

Figure 8. Radar charts – *Top:* typical radar chart - each apex of this polygon is a calibrated scale. *Middle:* types of radar charts with 3, 4 or 5 scales (red: important tests, blue: complementary tests, black: optional test). *Bottom:* Example of a sentinel coral host under heat/UV stress, using a 5-test grid on coral host, a 4-test grid on zooxanthellae photosymbionts and a 3-test grid on bacterial microbiome. Here, the coral holobiont undergoes preliminary signs of stress on photosymbiotic component (red star = oxidative stress on zooxanthellae).

STEP EIGHT (REPORT) - Health status of sentinel species and recommendations for amendment (Fig. 8 bottom)

Let us say we have a 7.4 average score on the coral host, 6.6 on the microbiome and 4.2 on the photosymbiont.. This indicates (i) what biological component is most affected by the environmental stress (ii) where appropriate action is to be taken. Here, holobiont looks normal but early signs of stress are detected on the zooxanthellae (ROS and oxidative stress products), without signs of microbial dysfunction.

On the basis of these visual diagrams and numerical scores that are explicit enough to be understood and followed by non-scientists. Useful recommendations can be made to custom- ers - if the principle behind this nicknamed "INDICORAL" environmental tool can be validated as a standard, legal enforcement can follow.

5.2. Strengths, limitations and future of environmental diagnosis tools

What is described here is a custom-designed diagnostic tool that integrates the critical snapshot information from different analytical approaches into a single easy-to-read layout, with a consensus fitness index. Using the multivariate "radar chart" model in conjunction with the fitness index then allows us to point out the weaknesses of the holobiont health at a given time, after which corrective measure can be proposed. Such a chart is made up of radiating spokes each representing a performance scale, for example rated from 0 to 10 in a given test. The spokes or performance scales correspond to a single test type, either molecular or visual or physiological. A robust radar chart typically integrates at least one test of each sort, in order to miss as little useful information as possible when establishing the final diagnosis. For example, the holobiont may "look normal" (as compared to control conspecifics) using a visual scale, perform close to optimal using a physiological assay, yet present strong signs of stress using an omics approach that detects early molecular responses. Or else, the holobiont may be suffering from a pollutant that acutely undermines its respiration, with no apparent effect on its microbiome nor on its appearance, and so on.

The objection that immediately comes to mind is that such an endeavour is time and money consuming, knowing how difficult it is to perform discriminating omics tests alone, especially when dealing with meta-data and the mathematical treatment that follows. The answer has to be optimistic: giant strides are being made in analyzing environmental meta-data more efficiently and cost-effectively, a good thing since some environmental issues are becoming rapidly critical. The other objection is that you need a whole team of specialists to devise and run a single "radar experiment".

The answer is two-fold: (i) once a suitable sentinel holobiont is chosen, the tedious part is to establish a reliable control database for each experiment type, mostly in the laboratory and using precise instrumentation. This database may initially require a panel of experts to set it up, but it does not have to be repeated for subsequent investigations. The other aspect is that we are creating a tool, not a thorough investigation into fundamentals, i.e. the experts must make sure that only essential information is retained, e.g. use identified molecular or microbial markers instead of profiles, set critical or threshold values within the 0-10 scale of responses (e.g. onset of symptoms appearance, loss of pigmentation, resilience limits, and so on), 0-2 representing the natural variability observed in controls, and 8-10 representing loss (no-return point to immediate death). This simplified representation relies on strong metrics, not on checking hypotheses which is the job of researchers. The custom association of 3 to 5 different tests types on the coral host and/or the photosymbionts and/or the mucus-bound bacterial flora spans a whole range of dysfunction possibilities.

5.3. Predicting biodiversity loss using the extended holobiont concept

The health status of the coral holobiont directly affects all associated life forms according to the degree of dependence on the host. Ultimately, the loss of the host determines the biodiversity loss of all dependant flora fauna and microbes. Changes in fish biodiversity and pressures on feeding niches have been reported in relation to coral bleaching and loss [51].

Pocillopora damicornis has been used as a choice model in biodiversity issues, not only because of its pan-tropical distribution in the Indo-Pacific region and because of its sensitivity to climatic stress, but also because it provides shelter to a number of commensals that are dependant to varying degrees on the fitness of the coral host [52] and on the state of its bacterial flora. Metagenomic barcoding of the whole extended holobiont system, coupled with appropriate bioinformatics, will be a powerful asset in determining biodiversity loss associated with environmental stressors, and provide a useful link with the analysis of detritic biota after the host's death.

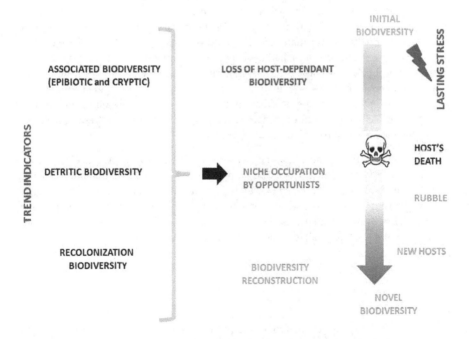

Figure 9. Loss of host-associated biodiversity (i) of original holobiont during lasting stress episode until host's death, (ii) detritic biodiversity finding refuge in coral rubble (iii) replacement biodiversity as substrate is recolonized with new hosts.

6. Measuring, predicting and hopefully mending

As pointed out earlier, over half a billion human live off goods and services of coral reef ecosystems at large, mostly in Asia. But not only: research on some 20,000 chemicals extracted from reef invertebrates has inspired the design of novel anticancer agents, antibiotics, anti-inflammatory, painkillers etc. and the exploration of the complex biosynthetic pathways leading to the production of these "miracle" molecules is only starting. Knowing the paramount importance of host-microbial symbioses in the making of these molecules, biodiversity loss will inevitably lead to chemodiversity loss, and opportunities will no longer exist to investigate the full bio-inspiration potential of what our holobiont systems can produce better than we can.

Marine ecology is a young science, and its developments are accelerating in response (i) to the urge of measuring the reactions of organisms facing the fluctuations of their environment, (ii) to the understanding of how they form complex interaction networks around key molecules acting as mediators of antibioses and symbioses, feeding hierarchies and occupational strategies for essential resources. Holobiont-wide systems biology is coming of age and will allow us to understand how various components of a holobiont system respond to stress in a coordinated manner, in the face of sudden and brutal environmental disasters, or of steadily increasing climatic or anthropogenic forcings, against the background of naturally fluctuating levels of stress. Ecosystem-wide resilience to environmental challenges can no longer be hoped for, and total biodiversity wipe-out of coral reefs is highly unlikely. The most probable scenario is that some reef systems might face near-total biodiversity collapse (e.g. in tropical zones that are within direct influence of urban expansion), while coral species from subtropical and remote localities will be constantly trying to acclimatize, each ecosystem striving to reach an overall "resilience equilibrium" at the cost of some of its biological diversity.

In 2013, most biological phenomena can be measured at all scales, from single cells to whole ecosystems, directly or using proxies and appropriate metrics. In combination, -omics, physiological/ecotoxicological and imaging tools provide a potentially formidable combination for measuring stress responses in coral holobionts and their separate components. Designing environmental tools such as proposed here might soon or later become a necessity, not only for coral reefs, but for all endangered ecosystems, marine and terrestrial.

Acknowledgements

I wish to dedicate this environmental tool concept to those who guided and encouraged my early endeavours as an enthusiastic observer of nature: to the great naturalist René Catala (1901 - 1987), and to my Australian and New Zealander academic mentors, from Sydney to Townsville via Auckland.

Author details

Stéphane La Barre[1,2]

1 Université Pierre et Marie Curie-Paris 6, UMR 7139 Végétaux marins et Biomolécules, Station Biologique F-29680, Roscoff, France

2 CNRS, UMR 7139 Végétaux marins et Biomolécules, Station Biologique F-29680, Roscoff, France

References

[1] La Barre S. Coral reef biodiversity in the face of climatic changes. Chapter 4 in: Biodiversity Loss in a Changing Planet, vol. 4 of Biodiversity series, Intech (ISBN 979-953-307-252-3), 77-112, http://www.intechopen.com/subject/biological-sciences/biodiversity (open access, Nov. 2011).

[2] Benton MJ. When life nearly died: the greatest mass extinction of all time. London: Thames & Hudson 2005 ISBN 0-500-28573 X.

[3] Knoll AH, Bambach RK, Canfield DE, Grotzinger JP (1996). Comparative Earth history and Late Permian mass extinction. Science 1996;273(5274): 452–457, ISSN 0036-8075

[4] Falkowski P. Tenth Annual Roger Revelle Commemorative Lecture: The once and future ocean. Oceanography 2009;22(2) 246-251, doi:10.5670/oceang.2009.57

[5] Vitousek PM, Mooney HA, Lubchenko J, Mellilo JM. Human domination of Earth's ecosystems. Science 1997;277(5325): 494-499. DOI: 10.1126/science.277.5325.494

[6] Connell JH. Diversity in tropical rain forests and coral reef high diversity of trees and corals is maintained only in a nonequilibrium state. Science 1978;199(4335): 1302-1310, ISSN: 0036-8075

[7] Baker B. New ocean policy depends on biological research. BioScience 2012;62(5) 524, doi: 10.1525/bio.2012.62.5.19

[8] Jackson JBC, Buss L. Allelopathy and spatial competition among coral reef invertebrates. PNAS 1975;72(12) 5160-5163, ISSN: 0027-8424

[9] Pawlik JR, Steindler L, Helkel TP, Beer S, Ilan M. Chemical warfare on coral reefs: sponge metabolites differentially affect coral symbiosis in situ. Limnol. Oceanogr. 2007;52 (2) 907-911, ISSN: 0024-3590

[10] Reshef I, Koren O, Loya Y, Zilber-Rosenberg I, Rosenberg E. The coral probiotic hypothesis. Environmental Microbiology 2006;8(2) 2067-2073, ISSN: 1462-2920

[11] Wilson EO, editor, Peter FM, associate editor, Biodiversity, National Academy Press, March 1988, 521 pages ISBN 0-309-03783-2 ; ISBN 0-309-03739-5

[12] Bonney R, Cooper CB, Dickinson J, Kelling S, Phillips T, Rosenberg KV, Shirk J. Citizen Science: a developing tool for expanding science knowledge and scientific literacy. BioScience 2009;59(11), 977-984, doi: 10.1525/bio.2009.59.11.9

[13] Reyers B, Polasky S, Tallis H, Mooney HA, Larigauderie A. Finding common ground for biodiversity and ecosystem services. BioScience 2012;62(5) 503-507, doi:10.1525/bio.2012.62.5.1

[14] Kelling S, Hochachka WM, Fink D, Riedewald M, Caruana R, Ballard G, Hooker G. Data-intensive science: a new paradigm for biodiversity studies. BioScience 2009;59(7) 613-620, doi: 10.1525/bio.2009.59.7.12

[15] Nichols JD, Cooch EG, Nichols JM, Sauer JR. Studying biodiversity: is a new paradigm really needed? BioScience 2012;62(5) 497-502, doi: 10.1525/bio.2012.62.5.11

[16] Rosenberg E, Koren O, Reshef L, Efrony R, Zilber-Rosenberg I. The role of microorganisms in coral health, disease and evolution. Nature Rev. Microbiol. 2007;5 355-362, doi:10.1038/nrmicro1635

[17] Grottoli AG, Rodrigues LJ, Palardy JE. Heterotrophic plasticity and resilience in bleached corals. Nature 2006;440, 1186-1189, doi:10.1038/nature04565

[18] Rosenberg E, Kushmaro A, Kramarsky-Winter E, Banin E, Loya Y.The role of microorganisms in coral bleaching.The ISME Journal 2009;3, 139-146, doi:10.1038/ismej.2008.104

[19] Rosenberg E, Zilber-Rosenberg I. Symbiosis and development: the hologenome concept. Birth Defects Research, part C 2011;93, 56-66, DOI: 10.1002/bdrc.20196

[20] Rohwer F, Seguritan V, Azam F, Knowlton N. Diversity and distribution of coral-associated bacteria. Mar. Ecol. Prog. Ser. 2002;243, 1-10, ISSN: 0171-8630

[21] Horgan RP, Kenny LC. Omic technologies: genomics, transcriptomics, proteomics and metabolomics. SAC review, The Obstetrician and Genaecologist 2011;13, 189-195, DOI: 10.1576/toag.13.3.189.27672

[22] Wang Z, Gerstein M, Snyder M. RNA-seq: a revolutionary tool for transcriptomics. Nature Reviews Genetics 2009;10, 57-63, doi:10.1038/nrg2484

[23] De Salvo MK et al. Differential gene expression during thermal stress and bleaching in Montastrea flaveolota. Molecular Ecology 2008;17(17), 3952-3971, DOI: 10.1111/j.1365-294X.2008.03879.x

[24] Traylor-Knowles N et al. Production of a reference transcriptome and transcriptomic database (PocilloporaBase) for thecauliflower coral, Pocillopora damicornis. BMC Genomics 2011;12:585, doi:10.1186/1471-2164-12-585

[25] Voolstra CR et al. Rapid evolution of coral proteins responsible for interaction with the environment. PLoS ONE 2011;6(5) e20392, doi: 10.1371/journal.pone.0020392

[26] Strychar KB, Coates MC, Sammarco PW, Piva TJ. Bleaching as a pathogenic response in scleractinian corals, evidenced by high concentrations of apoptotic and necrotic zooxanthellae. J. Exp. Mar. Biol. Ecol. 2004;304, pp. 99-121, http://dx.doi.org/10.1016/j.jembe.2003.11.023

[27] Strychar KB, Sammarco PW. Effets of heat stress on photopigments of zooxanthellae (Symbiodinium spp.) symbiotic with the corals Acropora hyacinthus, Porites solida and Favites complanata. International Journal of Biology, 2012;4(1) 3-19. DOI: 10.5539/ijb.v4n1p3

[28] Chow AM, Ferrier-Pagès C, Khalouei S, Reynaud S, Brown IA. Increased light intensity induces heat shock protein Hsp60 in coral species. Cell stress and Chaperones 2009;4, 469-476, doi: 10.1007/s12192-009-0100-6

[29] Mydlarz LD, Palmer CV. The presence of multiple phenoloxidases in Carribean reef-building corals. Comparative Biochemistry and Physiology, Part A. 2011;159, 372-378, http://dx.doi.org/10.1016/j.cbpa.2011.03.029

[30] Palmer CV, McGinty ES, Cummings DJ, Smith SM, Bartels E, Mydlarz LD. Patterns of ecological immunology: variation in the responses of Carribean corals to elevated temperature and a pathogen elicitor. The Journal of Experimental Biology 2011;214(24), 4220-4249, doi:10.1242/jeb.057349

[31] Vidal-Dupiol J et al. Innate immune responses of a scleractinian coral to vibriosis. Journal of Biological Chemistry 2011;286(25) 22688-22698, doi:10.1242/jeb.057349

[32] Martinez-Luis S, Ballesteros J, Gutiérrez M. Antibacterial constituents from Pseudoalteromonas sp. Rev. Latinoamer. Quim. 2011;39(1-2).75-83, ISSN: 0370-5943

[33] Kvennefors CE et al. Regulation of bacterial communities through antimicrobial activity by the coral holobiont. Microb. Ecol. 2012;63, 605-618, 10.1007/s00248-011-9946-0

[34] Shnit-Orland M, Sivan A, Kushmaro A. Antibacterial activity of Pseudoalteromonas in the coral holobiont. Microbiol. Ecol. 2012;64, 851-859, 10.1007/s00248-012-0086-y

[35] Rao D, Webb JS, Kjelleberg S. Microbial colonization and competition on the marine alga Ulva australis. Appl. Environ. Microbiol. 2006;72(8) 5547-5555, doi:10.1128/AEM.00449-06

[36] Franks A, Egan SH, Holmström C, James S, Lappin-Scott H, Kjelleberg S. Inhibition of fungal colonization by Pseudoalteromonas tunicata provides a competitive advantage during surface colonization. Appl. Environ. Microbiol. 2006;72(9),6079-6087, doi:10.1128/AEM.00559-06

[37] Miller DJ et al. The innate immune repertoire in Cnidaria – ancestral complexity and stochastic gene loss. Genome Biology 2007;8, R59, doi: 1186/gb-2007-8-4-r59

[38] Shinzato C et al. Using the Acropora digitifera genome to understand coral responses to environmental change. Nature 2011;476, 320-324, doi:10.1038/nature10249

[39] Shinzato C et al.The repertoire of chemical defense genes in the coral Acropora digitifera genome. Zoological Science 2012;29, 510-517, doi: http://dx.doi.org/10.2108/zsj.29.510

[40] Palmer CV, Traylor-Knowles N.Towards an integrated network of coral immune mechanisms. Proc. Roy. Soc. B – Biological Sciences 2012;279(4), 4106-4114, doi: 10.1098/rspb.2012.1477

[41] Kenkel CD et al. Development of gene expression markers of acute heat-light stress in reef-building corals of the genus Porites. PLoS ONE 2011;6(10), e26914, doi: 10.1371/journal.pone.0026914

[42] Vidal-Dupiol J, Ladrière O, Meistertzheim A-L, Fouré L, Adjeroud M, Mitta G. Physiological responses of the scleractinian coral Pocillopora damicornis to bacterial stress from Vibrio corallilyticus.The Journal of Experimental Biology 2011;214(9),1533-1545, doi:10.1242/jeb.053165

[43] Bayer T et al. Symbiodinium transcriptomes: genome insights into the dinoflagellate symbionts of reef-building corals. PLoS ONE 2012;7(4), e35269, doi: 10.1371/journal.pone.0035269

[44] Fabina NS, Putnam HM, Franklin EC, Stat M, Gates R. Transmission mode predicts specificity and interaction patterns in coral-Symbiodinium networks. PLoS ONE 2012;7(9), e44970, doi:10.1371/journal.pone.0044970

[45] Mouchka ME, Hewson I, Harvell D. Coral-associated bacterial assemblages: current knowledge and the potential for climate-driven impacts. Integrative and Comparative Biology 2010;50(4), 662-674, doi:10.1093/icb/icq061

[46] Barabábasi A-L, Oltvai ZN. Network biology: understanding the cell's functional organization. Nature Reviews – Genetics 2004(5), 101-113, doi:10.1038/nrg1272

[47] Bellantuono et al. Resistance to thermal stress in corals without changes in symbiont composition. Proc. Roy. Soc. - series B 2012;279, 1100-1107, doi: 10.1098/rspb.2011.1780

[48] Pett-Ridge J, Weber PK. NanoSIP: NanoSIMS Applications for Microbial Ecology. In: Microbial Systems Biology: Methods and Protocols. Methods in Molecular Biology, 2012;881, 375-408, DOI 10.1007/978-1-61779-827-6_13

[49] Palmer CV, Modi CK, Mydlarz LD. Coral Fluorescent Proteins as Antioxidants. PLoS ONE 2009;4(10): e7298, doi:10.1371/journal.pone.0007298

[50] Barott K, Smith J, Dinsdale E, Hatay M, Sandin S et al. Hyperspectral and physiological analyses of coral-algal interactions. PLoS ONE 2009;4(11): e8043, doi:10.1371/journal.pone.0008043

[51] Prachett MS, Hoey AS, Wilson SK, Messmer V, Graham AJ. Changes in biodiversity and functioning of reef fish assemblages following coral bleaching and coral loss. Diversity 2011;3, 424-452, ISSN : 1424-2818

[52] Plaisance L, Knowlton N, Paulay G, Meyer C. Reef-associated crustacean fauna: biodiversity estimates using semi-quantitative sampling and DNA barcoding. Coral Reefs 2009;28, 977-986, doi:10.1007/s00338-009-0543-3

Permissions

The contributors of this book come from diverse backgrounds, making this book a truly international effort. This book will bring forth new frontiers with its revolutionizing research information and detailed analysis of the nascent developments around the world.

We would like to thank Enrico Zambianchi, for lending his expertise to make the book truly unique. He has played a crucial role in the development of this book. Without his invaluable contribution this book wouldn't have been possible. He has made vital efforts to compile up to date information on the varied aspects of this subject to make this book a valuable addition to the collection of many professionals and students.

This book was conceptualized with the vision of imparting up-to-date information and advanced data in this field. To ensure the same, a matchless editorial board was set up. Every individual on the board went through rigorous rounds of assessment to prove their worth. After which they invested a large part of their time researching and compiling the most relevant data for our readers. Conferences and sessions were held from time to time between the editorial board and the contributing authors to present the data in the most comprehensible form. The editorial team has worked tirelessly to provide valuable and valid information to help people across the globe.

Every chapter published in this book has been scrutinized by our experts. Their significance has been extensively debated. The topics covered herein carry significant findings which will fuel the growth of the discipline. They may even be implemented as practical applications or may be referred to as a beginning point for another development. Chapters in this book were first published by InTech; hereby published with permission under the Creative Commons Attribution License or equivalent.

The editorial board has been involved in producing this book since its inception. They have spent rigorous hours researching and exploring the diverse topics which have resulted in the successful publishing of this book. They have passed on their knowledge of decades through this book. To expedite this challenging task, the publisher supported the team at every step. A small team of assistant editors was also appointed to further simplify the editing procedure and attain best results for the readers.

Our editorial team has been hand-picked from every corner of the world. Their multi-ethnicity adds dynamic inputs to the discussions which result in innovative outcomes. These outcomes are then further discussed with the researchers and contributors who give their valuable feedback and opinion regarding the same. The feedback is then collaborated with the researches and they are edited in a comprehensive manner to aid the understanding of the subject.

Apart from the editorial board, the designing team has also invested a significant amount of their time in understanding the subject and creating the most relevant covers. They scrutinized every image to scout for the most suitable representation of the subject and create an appropriate cover for the book.

The publishing team has been involved in this book since its early stages. They were actively engaged in every process, be it collecting the data, connecting with the contributors or procuring relevant information. The team has been an ardent support to the editorial, designing and production team. Their endless efforts to recruit the best for this project, has resulted in the accomplishment of this book. They are a veteran in the field of academics and their pool of knowledge is as vast as their experience in printing. Their expertise and guidance has proved useful at every step. Their uncompromising quality standards have made this book an exceptional effort. Their encouragement from time to time has been an inspiration for everyone.

The publisher and the editorial board hope that this book will prove to be a valuable piece of knowledge for researchers, students, practitioners and scholars across the globe.

List of Contributors

Evgenia Ryabenko
GEOMAR/Helmholtz Centre for Ocean Research Kiel, Kiel, Germany
Institute of Groundwater Ecology, Helmholtz Zentrum München – Germany Research Center for Environmental Health, München, Germany

Tsukasa Hokimoto
Graduate School of Mathematical Sciences, The University of Tokyo, Japan

Hubert Loisel
Laboratoire d'études en Géophysique et océanographie spatiales (LEGOS) CNRS: UMR5566 – IRD – CNES – Observatoire Midi-Pyrénées – INSU – Université Paul Sabatier - Toulouse, France
Laboratoire d'Océanologie et des Géosciences (LOG), INSU-CNRS, UMR 8187, Université Lille Nord de France, ULCO, Wimereux, France
Space Technology Institute (STI), Vietnam Academy of Science & Technology (VAST), Hanoi, Vietnam

Dinh Ngoc Dat
Space Technology Institute (STI), Vietnam Academy of Science & Technology (VAST), Hanoi, Vietnam

Vincent Vantrepotte and Cédric Jamet
Laboratoire d'Océanologie et des Géosciences (LOG), INSU-CNRS, UMR 8187, Université Lille Nord de France, ULCO, Wimereux, France

Qian Liao and Binbin Wang
Department of Civil Engineering and Mechanics, University of Wisconsin-Milwaukee, Wisconsin, USA

Stéphane La Barre
Université Pierre et Marie Curie-Paris 6, UMR 7139 Végétaux marins et Biomolécules, Station Biologique F-29680, Roscoff, France
CNRS, UMR 7139 Végétaux marins et Biomolécules, Station Biologique F-29680, Roscoff, France

Printed in the USA
CPSIA information can be obtained
at www.ICGtesting.com
JSHW011340221024
72173JS00003B/184